AF539439

UNDERSTANDING LABORATORY ANIMALS

UNDERSTANDING LABORATORY ANIMALS

By

Dr. Amita Sarkar

Dept. of Zoology

Agra College

Agra (U.P.)

(India)

DISCOVERY PUBLISHING HOUSE PVT. LTD.

NEW DELHI-110 002

First Published-2010

ISBN 978-81-8356-543-1

Published by:

DISCOVERY PUBLISHING HOUSE PVT. LTD.
4831/24, Ansari Road, Prahlad Street
Darya Ganj, New Delhi-110002 (India)
Phone: 23279245 • Fax: 91-11-23253475
E-mail: parul.wasan@gmail.com
info@discoverypublishinggroup.com
Website: www.discoverypublishinggroup.com

Printed at:

Sachin Printers
Delhi

Preface

The present title "Understanding Laboratory Animals" has been written for those students interested in careers in diverse fields of biological sciences. It provides a structured approach to learning by covering all the important topics in a uniform, systematic format. The book has been comprehensively designed incorporating recent advances in this fast moving field. It also provides accessible information on molecular biology in compact form for undergraduate students in biology and related life sciences. It is intelligible to the educated layman, though it deals with some complex ideas. It is an adequate text for all the requirements of students in this area. In addition, busy lecturers who require a quick reference compendium will find it useful, particularly for tutional planning. Simple, yet hopefully clear figures and tables are provided throughout the book.

The over-riding goal of this book, and indeed of the whole *Understanding series,* is to present the essential information concering molecular biology in a compact, readily accessible form which leads itself to student learning and revision. The convergence of various approaches has generated a rich panorama of detail, the significance of which we are still attempting to unraval. The present text has been written as an introduction to this rapidly growing field.

To make the work more comprehensive and informative, the author has consulted many authoritative books, research journals, abstracts, monographs etc., so there can be no claim to originality except in the manner of treatment.

The author expresses his thanks to his friends and colleagues whose continue inspirations have initiated him to bring out this book.

The author expresses his gratitude to Mr. Wasan and staff of M/s Discovery Publishing House Pvt. Ltd. for their whole hearted cooperation in the publication of this book.

In the mean time, the author will remain sincerely responsible for any shortcomings of the book and be grateful to the readers for their suggestions and constructive criticism for the continuous betterment of the book. He takes this opportunity to appeal to the readers to send their suggestions straightaway to his Publisher.

Author

Contents

1

EARTHWORMS

Earthworms are used in educational and research establishments and are also bred on a *commercial* basis to provide farmers with stock for certain kinds of land *improvements* involving, for example, the *production* of *humus*.

Some zoological societies have found the demand for worms so great that it has been necessary to breed them on a large scale to provide food for certain species of birds and *mammals*. They are also used for fish bait, fish food, and, when poisoned, as a bait to kill moles.

There are approximately 220 species of earthworms in the family *Lumbricidae,* of which about nineteen are common over much of *Europe* and have been spread by man to many parts of the world. These species have become the dominant *members* of the local *earthworm* fauna and have locally often replaced the *indigenous* species.

In Britain there are 25 species, but the most widely distributed belong to six genera: *Eiseniella* (4 species), *Allolobophora* (3 species), *Dendrobaena* (3 species), *Bimastus* (I species), *Octolasium* (2 species), *Lumbricus* (4 species). These species can all be found in a wide variety *of soils* and environmental conditions.

They differ from each other in habits, size and colour, and although there are certain endemic species which require specialized conditions of temperature, *humidity* and even soil acidity in order to thrive, many of the species in the genera mentioned can *nevertheless* be bred and *successfully* maintained in the *laboratory*.

ACCOMMODATION

Containers

Several types of *container* have been used with success for the culturing of worms, from *jam jars*, *specimen tubes*, *oil drums*, and *biscuit tins* to wooden boxes; the latter are *undoubtedly* the most practical and efficient.

When considering the type of container, it is important to bear in mind the necessity of providing sufficient aeration. *Soggy* and *waterlogged* areas must not be allowed to develop in the medium. This means, in fact, that if glass jars or tubes without *adequate drainage* are used the medium must be carefully turned over at intervals.

The size of box used is not important, but for convenience in *handling* and maximum production a good practical area is 46 *x 38* cm with a depth of 20 cm. The boxes should be made of wooden boards 1·5 cm thick, jointed and screwed together. *Angle* irons can be used to give additional support.

The wood can be preserved with either *Cuprinol* or *aluminium* paint. Drainage holes *3* mm in diameter at the rate of one per square decimetre should be *drilled* in the base of each box. For the efficient use of the available space the boxes may be stacked in tiers, from five non to eight boxes high.

The lowest box in each tier should be supported on a wooden base about 14 cm above the ground to allow correct drainage. The individual boxes in the tiers can be *separated* from each other by 5 × 5 cm blocks of wood running lengthwise. These blocks or *separators* make it easier to water the cultures.

The boxes need not be covered provided they are not overcrowded but, to avoid every possibility of escape, *wire-gauze* or *perforated-zinc* tops can be used. The surface of the culture medium can be covered with damp *squares* of *hessian*, but opinion is divided as to the wisdom of this.

Some workers maintain that this material helps to conserve *moisture*, keeping the surface of the compost dark and damp and encouraging maximum production of egg capsules, whereas others have found that besides twisting and wrinkling, the hessian tends to give an uneven distribution of water over the whole surface area.

Media for Box Production

The media should be kept at approximately pH 7, although slight acidity can be *tolerated* by some species.

Ordinary Soil

Ordinary soil is the most easily obtained culture medium and it can be used alone if necessary. *Sticky clays* and *heavy loams* should, however, be avoided. The soil should preferably be light and *sandy*, and all *lumps*, stones and *foreign bodies* must be removed.

It should be worked down finely by hand, spread out on shallow trays to a depth of about 2 cm, and allowed to dry in the air for some days before use. This helps to *eliminate* possible parasites and *predators*. It should then be *remoistened* until it contains about 30 per cent of water. The soil should never be cooked or baked.

Animal Manure

Dung may also be used alone or mixed with other substances. Any manure may be used, but horse or rabbit is recommended. This should be collected and dried out initially after which it is *subsequently remoistened* to give a water content of about 30 per cent.

Commercial Peat Moss

Commercial peat moss is easily obtained, but it is best used only as an additive to the media mentioned above. It does, however, have a high acidity (about pH 3.5), and so from 360 to 500 g of hydrated lime per 125 l must be worked in in order to give a pH of approximately 7 or 7.5.

Since it is dry on purchase it requires no preliminary drying out, for it contains no harmful organisms. The necessary moisture—content of approximately 30 per cent takes a little while to achieve, and the water should be added *gradually* in order to bring about an even distribution. It may be necessary to stand the wetted moss for some time in a *closely* packed heap.

Temperature

Normally no artificial heating is required except in cases where rapid *breeding* is desired. If necessary the cultures can be *warmed* by passing electrical heating wires through the bases of the boxes.

CULTURE

Breeding and Rearing Media

The recommended medium for breeding and rearing most species of earthworms is: 1 part of dung, 3 parts of soil, 5 parts of peat moss (treated with hydrated lime to give a pH of 7.0 or 7.5), and a small sprinkling of sharp sand.

The whole should be *moistened* to give a water—content of approximately 30 per cent. *Barrett* (1949) used the same *ingredients* for box culture, but varied the proportions and suggested a mixture of 1 part of manure, 1 part of screened *topsoil*, and 1 part of *agricultural* peat moss. To this mixture he usually worked in from 1 to 2 kg of corn meal per 125 l of finished compost. He states that corn meal has been found to favour the production of egg capsules.

Starting the Cultures

When ready, the boxes are filled to approximately 6 cm from the top with a mixture of the soil and treated peat moss. The dung or other food is placed in a layer over this. The next step is to stock the boxes with mature worms.

These may be purchased from a commercial worm farm or collected from the land surrounding the laboratory. If they are purchased, it should be possible to order preferred species but, of course, a random collection will most likely consist of a *heterogeneous* mixture of large and small species of *varying ages*.

The length of mature earthworms varies from approximately 17 mm in the case of *Bimastus* and *Dendrobaena* to 300 mm for *Lumbricus terrestris*. A mature worm can be recognized by the presence of the clitellum, a *glandular swelling* on certain of the anterior *segments* of the body. The clitellum may be solely on the *dorsal* side, or it may completely *surround* the body.

The position and number of segments contained by the clitellum is a character used in identifying the various species. The selected worms should be placed on the surface of the compost, and any which have not *burrowed* down within fifteen minutes should be removed as unsuitable.

The size of box recommended (base 46 × 38 cm, height 20 cm) should easily take up to 200 large worms such as *Lumbricus terrestris,* or up to 400 small ones such as *L. rubellus* or *Allolobophora caliginosa*. *As* a rough guide, large species need approximately 145 cm^3 of *compost* per worm, *small species* about half this volume.

Of course young specimens from a known stock can be reared in large numbers per unit; they must, however, be separated as their size increases with age.

Breeding

In nature, the periods of maximum activity and reproduction of worms are normally the spring and autumn. In the laboratory, on the other hand, the higher *temperature*, the maintenance of an *optimum* moisture content in the compost, and the presence of abundant food supplies allow breeding to continue over a longer period, and it is also more *rapid*.

Incubation of the eggs takes from 7 to 14 days in summer and up to a month in winter, whereas in nature it takes several weeks in summer and several months in winter. Smaller species, which take up to a year to mature in nature, will mature in 6 to 8 weeks in the laboratory, and the larger species will mature in the laboratory in 4 to 5 months.

The brandling, *Eiseniella foetida,* which is normally a fast-growing worm, only takes from 4 to 6 weeks in the *laboratory*. *Allolobophora* spp. have periods of diapause, during which they stop feeding and breeding and curl up in holes in the soil; this condition may last for from several days to many weeks, but normal activity is later resumed.

The diapause does not occur in *E. foetida*, nor in species of *Lumbricus*. Earthworms, and the *Oligochaeta* in general, are hermaphrodite, and cocoon production may begin as soon as the clitellum is fully developed. *Cross-fertilization* is normal, and the eggs are fertilized in the cocoon, a ring-like structure secreted by the *clitellar glands*.

The worm withdraws backwards through the ring and the eggs are released into the cocoon as it passes the opening of the *oviducts*. The *spermatozoa* obtained from the partner at mating, and subsequently stored, enter the *cocoon* as it passes the openings of the *spermathecae*.

The cocoon closes as the worm withdraws its body. In most species only one or two worms will develop from the eggs in a single cocoon, although up to seven have been recorded for the *brandling*, *Eiseniella foetida*.

Maintenance of the Cultures

Properly prepared cultures are not difficult to maintain, provided the moisture content is kept within the correct range. Laboratory cultures should be watered daily with the aid of a fire *sprinkler*, but *waterlogging* must be avoided. When watering it helps if the top 2 cm or so of compost is gently *raked* to allow easy *penetration*.

Soil cultures tend to dry out more rapidly than those which are based on a high proportion of organic material. It is also advisable to turn over at least the upper half of the culture *periodically* in order to facilitate a uniform *distribution* of *moisture* and an adequate exchange of gases.

Addition of Fresh Food

The *dung* or other food, such as corn meal or *chicken mash*, should be placed on the surface of the compost rather than mixed in with the culture. This will usually prevent the *worms* from working only at the *bottom* of the box, leaving the top layers *comparatively* untouched.

It is always better to regulate the food-supply and to avoid adding large amounts of soggy food at any one time. For up to 300 worms in peat moss, no more than about 40 g of air-dried dung should be added at any one time. Overfeeding of wet food leads to *fungal growths* which *spread out* and *spoil* the cultures.

Examination of Culture, and Collecting Cocoons

Cultures should be examined at least once a fortnight to see that they do not deteriorate, and also to gather *cocoons*. This gives an opportunity to check the water balance and to turn and aerate the culture.

If this is not done, the compost will become foul owing to the packing down of the medium and the growth of *undesirable fungi*. The soil or other compost should be changed every six months. For *examination*, the culture medium should be tipped out on to a tray to check the condition of the worms and to gather the *cocoons* for future breeding.

The intervals between *examinations* should at no time exceed three weeks, as that would give some cocoons a chance to hatch in the culture, and the resultant young would not then be recovered easily. The cocoons are roughly *spherical* in shape, and they vary in size according to the species.

In smaller species such as *Allolobophora caliginosa*, *A. chlorotica, Lumbricus rubellus,* and *Eiseniella foetida*, the cocoons vary in length from 2.5 mm to just over 3 mm, whereas those of *L. terrestris* are 6 to 7 mm long. The cocoons vary in colour from brown to orange and yellow.

Gathering the cocoons is rather a tedious process, but they can be easily seen and extracted quite cleanly. With cultures consisting mainly of soil, however, it may be necessary to wash the compost through a sieve (mesh 1.5 mm) with a jet of water, in order to get the cocoons clean.

If the washing method is used it is *advisable* to remove all the worms first, and then soak the culture for 24 hours before using the jet and sieve. If washing is carried out, new compost will be needed after each examination.

After the cocoons have been removed from the compost and counted, they should be carefully placed on very damp filter paper or *moist* sand and enclosed in a *glass* or *plastic* container where their progress can be observed; or, if desired, they can be placed out immediately into fresh *culture boxes*.

The cocoons are at first opaque but later become somewhat *transparent*, allowing the developing worms to be seen.

KILLING AND PRESERVATION

Earthworms can be killed with many chemicals. If dissection is not contemplated, then formalin is suitable as a preservative, Guild recommends a 4-per-cent solution initially, followed by transfer to a 5-per-cent solution after 2 to 3 days.

For immediate dissection 60-per-cent ethyl alcohol as a preservative allows worms to remain pliable for several months. If they are wanted for dissection after a longer period of time they should be preserved in 70-per-cent *alcohol*.

2

Frogs and Toads

Frogs and Toads have a number of uses in laboratories. They are in demand for teaching purposes in biological courses and may be required in simple routine anatomical dissections, for class experiments in the physiology of muscle and nerve and of cardiac function, or in extensions of this type of work for pharmacological purposes.

Females of *Xenopus laevis* and males of a large number of toads have been used for tests of human pregnancy, and similarly they can be used for the assay of gonadotrophins. Male frogs have also been used for this purpose, but since adrenaline produces the same spermiation response in these as gonadotrophin does the results are not completely reliable.

Changes in the skin colour of tree frogs have been shown to follow the use of ACTH under standard conditions; the response to standard doses has been followed and its sensitivity has been found to be very great. The response of individual melanophores in frog skin have been studied *in vitro,* and research has been carried out on the permeability of frog skin to various ions.

For toads there is an impressive volume of South American research on the processes of spermiation and ovulation, as well as on the secretions of the oviduct. Houssay used the giant species for much of his early work on the pituitary and pancreas.

Taylor and Ewer (1956) have investigated the sloughing of the skin and also the water balance in *Bufo regularis*. Laboratory workers have examined the processes of reproduction in a number of species, e.g. Savage (1934, 1937) in *Rana temporaria, Bufo bufo* and *Bombina bombina,* Mertens (1957) in *Gastrotheca ovifera,* and Amoroso *et al.* (1957) in *G.*

marsupiatum. Similar work is continuing on other species. If any accurate research is to be carried out, it is essential that this should be defined in terms of the species used, the size of the animals, and the conditions under which they have been housed and fed.

It has been found that the spermiation response of toads to gonadotrophin shows minute variations with different seasons even when the animals are kept under standard conditions, and differences in the temperature at which they are housed may cause major differences in response.

For work on changes in skin-colour, the animals must have been acclimatized over a period of time to the colour of their background. Healthy frogs and toads should be well nourished. This shows not in bodily bulk (which may vary with fluid content) but in plumpness of the thighs; wasting here is an indication of previous starvation.

The animals should be lively, and should have neither ulcerations nor abnormal reddening of the skin. Both of these are likely to be signs of the early stages of red-leg, a disease which is particularly fatal. Signs of lymphatic blockage or other infestation by parasites may be manifested by superficial oedema or by prominence of the lateral-line organs in *Xenopus*.

ENVIRONMENTAL REQUIREMENTS

The requirements, if housing of frogs and toads is to be *adequate*, depend upon certain of the *amphibians' characteristics*. The tail-less ones in general have moist skins through which water can pass in either direction, and they must therefore be protected from both water loss and the uptake of toxic substances through the skin.

While some species of *Salientia* have avoided the problem by spending their whole life in water (as in *Xenopus* and other genera), others have remained either within a frog's jump of the edge of a pond or in damp habitats, and similar conditions must be provided for them in the laboratory.

Yet others, as the toads and tree frogs, have adapted themselves to drier regions (although they also will become desiccated if kept too dry), while certain Australian species of waterholding frogs maintain their own damp environment internally during the dry season.

At first sight it is easy to suggest that only a few species should be kept under laboratory conditions, e.g. the common frog *(Rana temporaria)*, common toad *(Bufo bufo)* and the platanna or clawed toad

(Xenopus laevis). This, however, ignores the fact that in many countries a number of different species are used, while individual research workers may wish to investigate a number of problems involving such things as colour-change in tree frogs, the breeding behaviour of specialized forms (as in *Gastrotheca*, *Pipa,* or *Ascaphus*), or even the inheritance of colour pattern in frogs, all of which imply maintenance of numbers of amphibians over a considerable time.

As well as the variety of species, it is necessary to think about the separate problem of tadpoles. While biologists or anatomists may wish to rear numbers of tadpoles, which require conditions different from those for adults, few usually tackle the task of rearing these to adulthood, although it can be done in the laboratory.

In fact, where several generations are required, it *must* be undertaken. Stress should be laid upon the fact that unremitting care and attention will be required for such work. If this cannot be arranged, then it is better to liberate the newly metamorphosed froglets in a spot where there is some chance of catching them again when adult in two or three years' time.

The special problems involved over toxins and their absorption through the skin may be considered from various aspects. It must be remembered that the skin secretions of some *Salientia* are toxic to other species. This is not just a matter of the more exotic kinds such as the arrow-poison frogs of the genera *Dendrobates* and *Atelopus,* but of commoner species.

The skin secretions of the American *Rana palustris* are poisonous to other frogs, while common frogs *(R. temporaria)* kept for as little as half an hour in a sack with common toads *(Bufo bufo) will* be found dead at the end of the period. It is therefore common sense to keep such species apart, and only allow together those known to be mutually compatible.

Next comes the problem of death among the stock. Frogs are very sensitive to the products of decomposition, and a dead frog in the water container will soon result in other deaths among the community. It is therefore axiomatic that the cage be inspected daily and any dead or obviously diseased animals removed: this is of course normal practice in animal care, but cannot so easily be carried out when frogs and toads are in transit over long distances.

Finally, frogs may be sensitive to normal substances. Salt can readily kill them, although some species can survive in brackish water, while the crab-eating frog *(Rana cancrivora)* lives normally in sea water. The problem of metals is rather more complicated, and it has long been held that these are toxic to frogs and should therefore not be allowed in contact with them.

Work by Kaplan and Yoh (1961) has shown that solutions of more than 0·0015 per cent of copper sulphate are toxic to *Rana pipiens* although, earlier, Kaplan and Licht (1955) had recommended a thousand times this dose for an hour in treatment of the anuran bacterial infection red-leg, while for some years a prophylactic against this disease has been to leave a piece of copper wire in the water.

Elkan (1957) has advocated the use of galvanized metal tanks for *Xenopus laevis*. In general it is usually considered wise to keep amphibians in containers where they are not in direct contact with metallic surfaces.

IDENTIFICATION OF SEX

Sex differences vary from species to species. In European frogs and toads the male has black swellings on the thumb and first two fingers, as well as thicker forearms than the female, In *Pelobates* these nuptial pads are found farther up the forearms.

Such nuptial pads are most apparent in the breeding season, and practice is required to sex *B. bufo* at other times by noticing the rudimentary pads then visible. In the tree-frog *(Hyla arborea)* there are no pads, and the male is distinguished by the discoloured underside of the throat caused by the deflated vocal sacs.

The blackish throat is the distinguishing mark of the male of *Bufo regularis*. In *Xenopus* species the distinction between the sexes lies in the small labial folds found by the cloaca of the female. It is apparent that there is great variation in the sexual distinctions among different species; the only way of making sure that the right sex has been supplied is to know the peculiarities of the species being used.

BREEDING AND REARING

Breeding of frogs and toads is most readily carried out with freshly caught individuals, which in many cases have come straight from the breeding pond. Mating and oviposition will occur if these are kept under suitable conditions, which depend in turn on the species involved. In general the eggs are usually laid in water where they hatch into free-swimming tadpoles.

Tadpoles have requirements different from those of adult amphibians, which in some ways are easier to satisfy and in others more difficult. They usually need to live in water, which must be sufficiently aerated, contain adequate food, and yet be kept free from decaying food particles and faecal matter.

These requirements are best met by maintenance under aquarium conditions, a small pump being used if necessary to aerate the water. Conditions inside the aquarium will depend on the species concerned; some survive well in clear water, while other tadpoles prefer weed for shelter.

Some live harmoniously together, while others only do so when food is plentiful and may, in times of scarcity, turn to cannibalism, which in turn others may practise regularly. Species like *Rana ridibunda* which feed voraciously on other tadpoles must therefore be kept isolated in small individual containers.

Feeding tadpoles is relatively easy. They will thrive during their early stages on vegetation or on the diatoms found on this. This diet may be natural, in the form of waterweed or algae, or alternatively a substitute such as a piece of lettuce leaf may be left to float on the surface.

Under these conditions little cleaning of the tank is needed. However, the latter stages of tadpole life, with the appearance of the limbs and regression of the tail, require a protein diet. This can be provided by giving small pieces of meat, but equally good are small amounts of high-protein artificial foods, which can be sprinkled on the surface.

When the tadpoles have almost lost their tails they will shortly be living a terrestrial life, so access to land must be provided. If the tank has no land surface, the water level may be dropped to a mere 5 cm and a couple of bricks be placed on the bottom.

Once the froglets or toadlets have metamorphosed completely the phase of major difficulty is reached. Unless there are strong reasons for rearing them further, it is advisable now to release the young. Lees (1962) has recently described a method of raising them in special outdoor frog pits covered with nylon net, *Drosophila* and aphids being used as food.

He points out that the provision of an excess of such food seems to be the necessary stimulus for regular feeding in captivity. Frogs have also been reared under vivarium conditions, when it is possible to start them on house-flies almost immediately after metamorphosis

Under these conditions Green (personal communication) has found that they will not grow on such small fry as *Drosophila* and aphids alone, but that house-flies also must be given from an early date.

Possibly the divergence of opinion here relates to the total bulk of food supplied.

In the case of aquatic species such as *Xenopus laevis,* the baby platannas can be fed on *Daphnia* and then on bloodworms and meat. It must be remembered that Bruce and Parkes (1950) showed that young *Xenopus* reared even on horse liver develop severe rickets, so that rat or rabbit liver must be given at regular intervals if this is to be avoided.

HANDLING

Frogs and toads should be clasped by the hind legs, or (with less assurance) by the body. If the skin is moist or slippery, a rough cloth or duster should be used. Handling for pregnancy tests is described in a later section.

ACCOMMODATION

The type of container required for amphibians will depend upon the species, the number to be housed, and the period for which they are to be kept. Over a short period a relatively small container will suffice for a small number of frogs or toads, but a period of a week or more requires more specialized conditions, which will vary according to whether the animals are being kept at room temperature, under artificial hibernation, or at tropical temperatures.

At Room Temperature

Frogs

These are normally kept in large sinks from which slowly running water removes any toxic excreta. Galvanized-iron or stainless-steel tanks or glazed porcelain sinks may equally well be used. Bricks rising above the surface of the water give footing for the frogs, and damp moss can supply them with shelter.

Alternatively they may be kept in aquaria where the water is changed every few days. Whatever the container, it must have a close-fitting lid of either wire mesh, expanded metal, or perforated zinc. The holes in this should be of such a size that flies just cannot escape through the mesh, while the metal remains as strong as possible.

A hole 2.5 to 4 cm in diameter should be cut in the lid and kept plugged with a cork; this can then be used for introducing bluebottles, other flies, or moths as food.

When shy species such as *Rana esculenta* are kept, the sink must contain some deep water as well as land, and must be in a room remote from disturbance. Without these conditions the frogs will not feed, and will therefore inevitably become emaciated.

Toads

In this classification are included for present purposes not only members of the genus *Bufo* but also certain other species with terrestrial habits and a dry skin, which do not require such damp habitats as frogs do.

Similar types of container may be used for them, but when any numbers are kept it is convenient to have special containers constructed on the plan of giant vivaria. These may be made of galvanized metal, about 4 ft (120 cm) long by 21 ft (75 cm) deep, the front being about 6 in (15 cm) and the back about 9 in (23 cm) high.

The side walls then vary from 23 cm at the back to 15 cm at the front. In one of these side walls there is a hole about 11 in (4 cm) in diameter, and this is normally kept closed either by a cork or by a metal plate pivoting on a pin, the position of equilibrium being the closed one.

This hole is for the introduction of both food and individual amphibians on occasions when it is not desirable to lift the lid. The vivarium is roofed with two slabs of 1-in (67mm) plate glass, and a ledge in (6·3 mm) wide must run round the inside of the vivarium, 1 in (12.5 mm) from the top.

This ledge must be made with care, so that when the lid is on there shall be no gaps through which small invertebrates could escape. Ventilation holes 0·15 cm (0·055 in) in diameter should be drilled at 1-in (2.5-cm) intervals along the sides, roughly 1 in (2.5 cm) below the ledge. Once it is made, the vivarium should be stoveenamelled on the inside.

The two slabs of 4-in (6-mm) plate glass, which roof the vivarium, must be cut to fit exactly. After being cut, the edges are ground smooth and two holes are drilled in each slab, roughly 6 in (15 cm) from the front and back edges and midway between the two sides. In these holes are bolted handles of a suitable size for lifting the lid.

It may be desirable to replace one of the two lids by a similar one of perforated zinc or expanded metal, and in this case the usual precautions must be taken over the size of the holes if food species are not to escape. Such a lid has an advantage over a completely glass one in allowing water vapour to pass through it, so that the humidity does not become too high.

Vivarium-type cages may be kept in racks or on shelves, or it may be more convenient to use a low trolley with several shelves for them. This can be made in skeleton form for both lightness and ease

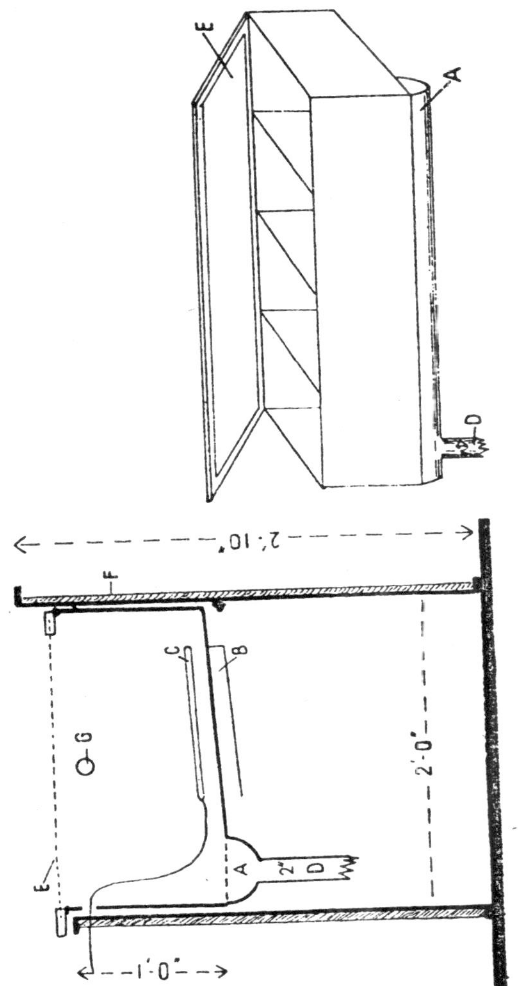

Figure 2.1: Galvanized-steel tank for Xenopus. (a) Section seen from the right in (b); (b) tank with asbestos F and support removed. A, Gulley with perforated tray to collect dirt; B, pocket for heating element of 100 to 200 W; C, phial for distant-control thermometer; D, drain pipe, 2 in in diameter, with full-gate valve; E, lid fitted with windolite; F, asbestos millboard, insulating the tank; G, water inlet.

of handling the cages and their inmates. For various species of toads the floor of the cage may be covered with earth or sand, but difficulties associated with cleaning may then appear.

Cover can best be provided by *placing* moss all over the floor of the cage, of sufficient depth for the toads to hide in it. Shelter may also be provided in the shape of broken flowerpots, laid on their side. In all cages there should be a supply of water.

This can either be in a large enamel dish, or in a big, stable earthenware dog bowl. The water level must not be such that toads can drown before they find their way out. Cages have not yet been designed where the water is provided in a basement storey with a ramp for ingress and egress, but this seems an obviously useful future development towards saving space.

Tree-Frogs

Although these are more akin to toads than to frogs, they require somewhat damper conditions. It is most convenient to keep them either in an aquarium with about 6 cm of water, roofed with perforated zinc, or else in the cylindrical type of caterpillar cage.

This has a round tin for base, topped with a cylinder of transparent plastic and bearing a perforated lid above this. As soon as obtained, such a cage should receive an extra coat of paint which is left to dry before 2 to 3 cm of water is placed in it.

Either type of cage can be furnished with a spray of foliage if desired, and the lid must have a hole cut, plugged as usual with a cork, for the introduction of flying insects as food. The tree-frogs will remain either on the foliage or on the vertical walls of the cage.

Platannas

The common type here is *Xenopus laevis,* but other species of *Xenopus* may also be kept similarly, either without special heating or at warmer temperatures. *Xenopus laevis* will survive at normal room temperature, so that heating need not be provided except in very cold weather. Any kind of aquarium tank may be used.

Elkan (1957) has described a galvanized steel tank approximately 8 × 4 ft (2.4 × 1.2 m), complete with heating element and thermostat, and with a sloping floor so that remnants of food can easily be drained out. Tadpoles of this species can easily be kept in an ordinary laboratory sink.

Other Species

A variety of other species can be accommodated in a wide range of the cages or tanks already described, depending on their requirements. Thus, for example:

Gastrotheca marsupiatum

These require fairly dry surroundings and can tolerate great heat. If breeding is to take place, then shallow water only is required for the mother to deposit the tadpoles, it being remembered that the adult is easily drowned in deep water.

Pipa pipa

A tank similar to that for *Xenopus* is required. The needs for breeding in captivity have not yet been elucidated, but these may perhaps include muddy water, decayed leaves and a particular pH of the water. Otherwise, feeding and care is as for *Xenopus,* but at a higher temperature.

Australian species from arid areas

Although these frequently remain buried in the ground outside the wet season, they have been found to survive captivity well at room temperature if kept in half an inch (12.5 mm) of water with some broken flowerpots for shelter.

Artificial Hibernation

The first requirement for this is that the frogs or toads should be in good condition and well fed. Those received from dealers in spring, therefore, which have just come out of hibernation before being collected from their breeding ponds, are not suitable for immediate cooling.

If it is intended that they be used for this, they must first be well fed under normal conditions, and then left unfed for a few days. Frogs can be thereafter kept in the normal laboratory refrigerator; preliminary investigations have shown that deaths will occur if the shallow water in their containers is not changed regularly.

Allison (1957) has described a successful technique of artificial hibernation, using trays 3 to 4 in (8 to 10 cm) deep, covered with fine-mesh wire netting, and she has kept toads similarly but with slightly damp moss rather than water. She has pointed out that while frogs will tolerate a constant temperature of 3°C (38°F), toads should not be kept below 4 to 7°C (39 to 45°F).

So far as is known, no experiments have yet been carried out to see whether it is possible to keep frogs under the usual sink conditions in a cold room.

Temperatures Above Normal Room Temperature

Where tropical species require to remain under warm conditions, it is frequently best to keep them in a thermostatically controlled room. Where only a few cages need such special treatment, or where small numbers require to be at different temperatures, it is often most convenient to heat cages individually.

For a narrow range of temperature, thermostatic control is normally best. In the case of aquatic species, an electric heater of the normal aquarium type can be used. Where a wider range can be tolerated, it is sometimes possible to use a carefully adjusted spirit or paraffin lamp to heat the base of the cage, provided this is made of a suitable material.

NUTRITION

The findings of Honigmann (1944-5) showed that the Common Toad *(B. bufo)* tends to eat anything of a suitable size which moves in relation to its background. Following on this, it is frequently felt that frogs and toads will eat any invertebrate or vertebrate of suitable size.

On the other hand Cott (1950) has shown from stomach contents that, in the wild, particular species tend to eat particular kinds of invertebrates, while Frazer and Rothschild (1961) have noted that a feeder even as voracious as *B. bufo* may reject distasteful insects.

Further, their results show that species distasteful to one vertebrate species may be perfectly palatable to another: this probably holds good for amphibians.

In practice a variety of food, both living and dead, may be given:

(1) Purely aquatic species such as *Xenopus will* take fresh pieces of meat, rat liver, placentae, foetuses and living invertebrates or even small fish.

(2) Other species will take a variety of suitable live foods, for example

Bluebottles

These may be reared from maggots. The most convenient way to do this has been found to be by leaving the maggots in fresh sawdust to pupate. This can be in some container such as a conical flask, but under these conditions moisture condenses in the flask, so that not only is wing-development prevented but the flies themselves tend to be stuck to the walls of the container and cannot easily be fed to the amphibians.

When flies are reared in bulk, it is found convenient to place the maggots and sawdust in a large biscuit tin from which most of the centre of the lid has been removed and replaced by perforated zinc of suitable mesh, soldered in.

A hole on one side, near the top of the tin, should be fitted with a piece of glass tubing which is plugged with cotton-wool; alternatively the tube can be of Perspex and fitted with a Perspex slide to close it.

The edges of the lid must be sealed to the tin with adhesive tape to prevent the egress of newly emerged flies. When flies are seen to have emerged in the tin, they can be fed to the amphibians by inserting

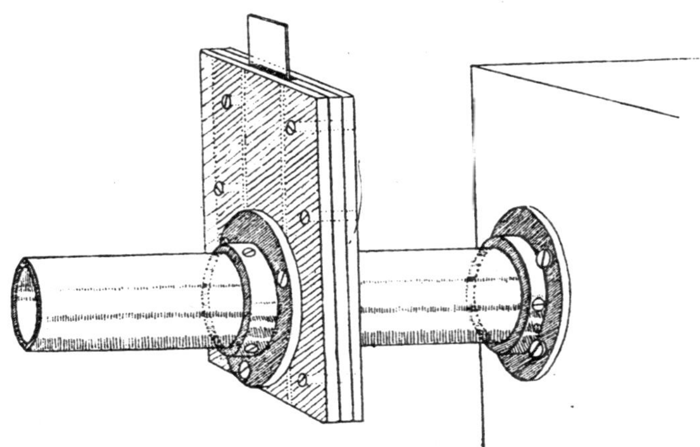

Figure 2.2: Perspex tube for admitting emergent flies from emergence cage into amphibians' vivarium.

the tube through the hole in the cage side or roof, and then removing the plug or slide which blocks it. A piece of wood or cardboard is placed over the lid of the fly-box, and the flies then move towards the light and emerge into the cage.

It is inadvisable to leave the fly-box and cage permanently together, since toads have been known under these conditions to penetrate the fly-box and die there.

Bluebottles and to a lesser extent house-flies are a very useful standby, and certain species have survived for years on these alone. In the case of toads it is advisable to alternate flies with maggots, and an excess of either must be avoided.

B. bufo in particular will pass undigested maggots if too many are given, while excess of bluebottles in the cage will cause panic, and dead toads may be found there in the morning. An excess of bluebottles may be fed, however, provided that the moss is deep enough for the toads to take cover in it.

In contrast, the Green Toad *(B. viridis)* is not affected by bluebottles in this way, and will merely continue eating until all the food is gone.

Mixed Insects

A varied diet is of greater value than one based on a single species. Many moths and other insects may be obtained by the use of a light-trap, and form a very valuable source of food for most species. Alternatively, sweeping a net over herbage or beating foliage over a sheet will yield a good haul which will be very useful.

Locusts

These can be very useful for any species large enough to take them (e.g. *Bufo marinus, B. arenarum* and *B. ictericus),* but use of the young stages to feed smaller species, while possible, is rather wasteful of the locust stock.

Other Invertebrates

Ants have been shown by Cott (1940) to be particularly common in the stomach contents of *B. bufo,* but as a live food they find escape too easy to be of much use. When a mixed diet is given, spiders and woodlice prove particularly useful.

They may be collected like insects or by searching under stones, beneath bark, under logs and the like. Worms are a great standby. As in the case of woodlice and other secretive and wingless species, it is advisable to give them in a special feeding cage which is kept free of soil or vegetation in which they might hide, since the amphibians will not then search for them.

As worms will drown in water, care must be taken to search and to remove any corpses if they are given to frogs or *Xenopus.* Whiteworms *(Enchytraeids)* can be given to small species like the Fire-bellied Toad *(Bombina bombina).*

Allison (1957) has listed a number of food species together with their availability and use. Terrestrial species will also take raw meat if this is of suitable size and is moved suitably, e.g. by a fine wire. A more satisfactory way of giving this is by forced feeding; see the following paragraph.

Artificial diets have been recommended; they are made up to a firm consistency and portions are force-fed twice a week to individual frogs. The technique for this comprises holding the frog in one hand, the thumb being on one side while on the other side the middle finger and third finger are on either side of the frog's forelimb, clasping it firmly.

With the free hand the edge of a spatula is gently inserted between the frog's jaws. Once the jaws are open, the forefinger of the other hand is used to keep them open, being placed in the angle of the jaw.

The free hand is then used to insert the food, and the frog's mouth is held shut and the throat is massaged until the frog swallows. By expending some of a technician's time it is thus possible to ensure that all the laboratory stock of frogs are kept in reasonably fed condition.

WATER

This is required by amphibians rather as a part of their environment than as a drink. Absorption through the skin can fulfil all their needs (see above) provided that the water is not harmful through being contaminated with stale food or corpses.

LABORATORY TECHNIQUES

The vast majority of laboratory amphibians are used as a source of anatomical material or physiological tissues. Those used alive for experimental purposes are few, and by far the greatest number of these are employed in pregnancy tests.

The experimental requirements will vary with the species. For *Xenopus* females a large jar will take an individual in 3 or 4 in (8 or 10 cm) of water, with a grating on corks below that, so that any eggs produced can drop through this without being eaten.

A pair of male toads or tree-frogs can be placed in a similar jar or a 400-ml beaker covered with a tile and containing approximately 1 ml of water. If several tree-frogs are required a larger beaker should be used.

Injection of fluids is generally into the dorsal or ventral lymph sac; some experts advise that the hypodermic needle should be introduced through the thigh muscles in order to prevent leakage. If this technique is not employed, a long fine needle (No. 12 or 17), already attached to the syringe, should be inserted about the middle of the back, the point being passed back as far as possible towards the tip of the urostyle.

The injection is made slowly in order to prevent reflux of fluid. For sampling cloacal urine, the toad or tree-frog is held by the body with the hind legs flexed, a fine glass pipette (1 mm or less in external diameter) is inserted gently into the cloaca for a short distance.

Urine will often track up this at once but, if this does not occur, gentle manipulation of the pipette backwards and forwards will soon produce it. The need for gentleness in this technique must be stressed.

RESTRAINT AND ANAESTHESIA

The only restraint normally used for the tailless amphibians is that involved in handling them. This has already been considered.

Anaesthesia is seldom needed except for short demonstrations of the capillary circulation. For this, the frog is immersed in a 1-per-cent solution of ethyl carbamate (urethane) until anaesthetized.

This solution can be used for recovery experiments, the frog being placed for the recovery phase under a dripping tap, when the ethyl carbamate becomes dialyzed through the frog's skin.

The same technique can sometimes be used with success for recovery of toads which have been injected with pregnancy urine, when the sample proves to have been contaminated with sedatives or other toxins.

As ethyl carbamate is carcinogenic it may be preferable to use tricaine methanesulphonate (MS222 Sandoz). Ahrenfeldt (1953) has tested the use on frogs and toads of hexylresorcinol, but has pointed out the narrow safety margin in this case between anaesthesia and death.

Euthanasia can be carried out with an overdose of anaesthetic or by holding the amphibian by the hind legs and tapping the back of the head sharply on the edge of a sink. This can then be followed by decapitation and/or pithing, as required.

TRANSPORT

Within the laboratory, amphibians may be transported in large earthenware or glass jars. For the more active species, such as tree-frogs, a lid is essential. The more lethargic ones, such as toads, may be carried in buckets.

By road, rail or air they travel best in closed boxes or cans. No food is required during such journeys. If a long, slow trip is envisaged, arrangements should be made for pouring in a little water occasionally.

Tree-frogs have travelled well by air from as far as Australia in small tins with a few holes in the lid, containing some damp cloth, and with a few pieces of twig wedged across to afford foothold to the inmates.

If holes are punched in the tin lid these must have the sharp protuberances on the outside. On one occasion, toads travelling on a relatively short journey by passenger plane died mysteriously when half-way to their destination, after leaving the previous stop alive and healthy.

Although no insecticide could be detected in their bodies, it is suspected that they may have been the victims of a spray which reached them through the open mesh of the lid of their box. While frogs and toads normally travel by air freight as livestock, if special arrangements are made as in this case for them to travel by another type of aircraft, every endeavour should be made to see that their box is airtight enough, and that the pilot fully understands the perishable nature of his passengers.

HEALTH

Diseases

There are no notifiable diseases carried by frogs or toads. The common ailment which is most destructive to stocks is the infectious disease red-leg, caused by *Proteus hydrofcls* and Borden, 1942) and its allies, or sometimes *Bacterium alkaligenes*.

A number of parasitic worms and protozoa may be found in the cloacal urine, and become quite plentiful in toads used for pregnancy tests several times during the summer.

Bieniarz (1950) has pointed out that encysted forms of protozoa only occur in the cloacal urine of non-breeding frogs after treatment with human pregnancy urine or gonadotrophin, which appears to cause a change from the asexual to the sexual mode of reproduction.

Thus, amphibians' own gonadotrophins as well as any given to them experimentally may affect the parasites present. Lees and Bass (1960) have shown that oestradiol in oil solution apparently lowered the level of parasitization of frogs by helminths.

Lees (1962) has pointed out that freshly purchased frogs often contain 20 to 30 lung flukes *(Haplometra cylindracea)* as well as the intestinal ones *Dolichosaccus rastellus* and *Pleurogenes claviger*.

Elkan (1957) has noted the trematode *Diplostomulum xenopi* in the pericardial sac of *Xenopus,* where he states that Nigretta and Coates found large numbers of active larvae and concluded that cardiac embarrassment led to the death of the host.

Elkan (1960) has listed five species of worms commonly found in laboratory *B. bufo,* but considers that only two of these, *Acanthocephalus ranae* and the tapeworm *Nemototaenia dispar,* may be dangerous parasites, the others (like many of the invertebrates mentioned by Noble, 1931) being harmless commensals.

Lees (1962) has mentioned the nematodes *Rhabdias bufonis* as a lung parasite in the lung and *Oswaldocruzia filiformis* in the intestine.

Elkan (1957) has described in *Xenopus a* spastic paralysis of the hind legs, which may extend to affect the forelimbs. If the animal is prevented from drowning by being kept in a shallow dish and fed, it will often recover spontaneously.

First Aid

First aid is normally required for amphibians only when these are suffering from the effects of toxins. All that can be done for them is to leave them under a running tap, while if there is associated oedoma a syringe should be used to draw off the excess fluid from the lymph sac.

Prevention and Control of Disease

Prevention is a matter of quarantining newly arrived stock for a week or two, with daily inspection for red-leg. During this period it may pay to keep them in 0.15-per-cent saline solution, to leave a piece of copper wire in the water or to sprinkle sulphadiazine powder on the surface of this. Any infectious disease should become apparent during the quarantine period.

In the case of young growing animals, the prevention of rickets (as in *Xenopus laevis*see above) depends upon a proper diet. Apart from red-leg and similar infectious diseases, this is the only one which merits consideration.

The question of curing disease in frogs and toads calls for setting the cost of the cure against the value of the victims. With common species it is frequently better to kill those affected and isolate their cage mates in quarantine against any further attacks, but there are occasions when every effort should be made for a cure.

In red-leg, it is essential to control the infection, but the sensitivity of the frog to drugs is one difficulty. Kaplan and Licht (1955) have recommended exposure of frogs to 1 2-percent copper sulphate for at least an hour, while others believe in the injection of streptomycin; Miles (1950) recommended a total of 1,000 units given subcutaneously over nine days, and found that this was accompanied by the recovery of frogs treated in the early stages of the infection, while it delayed the death of one with the disease in a more advanced stage.

Smith (1950) has cured *Bufo marinus* by giving 5 mg per 100 g of body-weight of chloromycetin by stomach tube, and following this up by 3 mg/100 g twice daily for five days. Hunsaker and Potter (1960) used the same treatment for a mixture of fifteen species of toads: this was successful at first, but a chloromycetin-resistant strain of the infective organism appeared. The disease was then successfully treated by giving 10 mg/g of achromycin V intraperitoneally, followed by twice-daily doses of 5 mg/g for a week.

Kaplan has suggested that 0·05-per-cent copper sulphate applied for a matter of minutes could be used to treat *Saprolegnia* infection.

3

THE GUINEA-PIG OR CAVY

Guinea-Pigs are natives of South America where their wild ancestors *(C. cutleri)* are still to be found in Peru. The date of their introduction into Europe and other parts of the world is not known. '*Guinea*' may have been derived from '*Guiana*', or it may merely have meant 'foreign'.

In the wild state guinea-pigs are partly diurnal and live in burrows. They are vegetarians, and the noises they make have been described as faint squeaks and grunts. Over the years, guinea-pigs have been kept as pets and a considerable fancy has developed.

Three main varieties are recognized, viz. the English. the Abyssinian and the Peruvian. The first is the commonest, is short-haired (1½ in, 3.8 cm) and occurs either as self-coloured (white, black, agouti, sandy. red, chocolate, cream, etc.). as a combination of two colours ('*bicolour*') or of three colours ('*tricolour*').

The colouring may be similar in pattern to that recognized by rabbit-breeders, e.g. Himalayan or Dutch, and the fancy recognizes such descriptions. The coat of the Abyssinian is short but rough and radiates as rosettes from a number of centres. This variety also occurs in a number of colours.

The Peruvian variety is long-haired (6 in, 15 cm) and is the least common. Guinea-pigs have long been used for animal experimentation because they are small, tame animals which are easy to control. The English variety of guinea-pig is more suitable than either the Abyssinian or the Peruvian.

In Great Britain the white-and-cream strain established by Dunkin and Hartley in 1926 has been widely adopted as a standard laboratory strain, but most of the larger populations are now composed entirely of pink-eyed white animals.

Of recent years this strain has been imported into many countries, and substrains are now well established, e.g. the Jap-Hartley strain at the Walter Reed Army Institute of Research in Washington. Other strains, whose origins and history are known, are listed in the *Catalogue of Uniform Strains of Laboratory Animals Maintained in Great Britain* published by the Laboratory Animals Centre, Carshalton, Surrey, England and in the *Handbook of Laboratory Animals* published by the Institute of Animal Resources (NAS-NRC), Washington D.C., U.S.A.

In health, the English guinea-pig is an alert animal; its coat is smooth, shiny, and unsoiled by faeces and urine. When handled, its coat should be dense and the hair strong but not harsh; it should be well fleshed and should not feel '*light*'.

There should not be any discharge from either the nose or the ears, nor should there be any evidence of slobbering or diarrhoea. Its breathing should be regular and noiseless and its posture and gait should be normal. Within a single strain there should be little variation in size or weight for a given age.

The normal guinea-pig has a pulse rate of 150 to 160 per minute, a respiratory rate of 110 to 150 per minute, and a rectal temperature between 39 and 40°C (102.2 and 104°F)

The red-cell count lies between 4.5 million and 5.5 million per μl, and the white cells number between 4,000 and 7,000 with an average of 5,800 per μl.

The percentage composition of the white-cell count is highly variable in different animals. The lymphocytes on average account for half the cells, but vary from 29 to 74 per cent, and the other principal cell, the neutrophile, may vary between 20 and 60 per cent, averaging 40 per cent.

Eosinophils basophils and monocytes in varying small numbers are present also. The weight of a guinea-pig at birth depends partly upon the level of nutrition in the colony and partly upon the number of young in the litter. The birth weight of individual animals varies considerably.

Heaviest weights often approaching 150 g are recorded for single births, but a fair average in litters of three or four youngsters is 90 g for males and 85 for females. Youngsters weighing less than 50 g at birth frequently fail to survive.

The average number of young per litter under good management should not be less than 3.0. but some breeders achieve 3.5 and specially selected stocks have maintained an average of 4.0 over periods as long as one year. In the herd supervised by the writer, an unselected group of 100 newly-born males, averaged 94.8 g, and 100 females averaged 91.9 g. At weaning 21 days later, the males averaged 248.8 g and the females 2402 g.

The mothers of these youngsters littered down in individual boxes and the average number of young per litter was 3.6. Development is rapid, and young stock should increase in weight at around 4 to 5 g a day during the first two months, at the end of which time they should weigh 350 to 400 g.

The growth rate then slows regularly and guinea-pigs reach full maturity at a weight of 750 g (males) and 700 g (females) during their fifth month, but continue to grow slowly until they reach the age of 15 months. Males may reach 1,000 g and females about 850 g.

The principles involved in the rearing and keeping of guinea-pigs are similar to those which govern the husbandry of other laboratory animals. The aim must always be to produce guinea-pigs free from defects which will invalidate the results of experiments, and at the same time to promote their comfort and well-being.

HUSBANDARY

Breeding

Foundation stock should be obtained from an existing herd which has a satisfactory history regarding both health and production. The animals should be obtained soon after weaning and placed immediately in the quarters where they are to breed, with their own attendant and properly isolated from any other stocks.

By obtaining them at this early age it will be possible to hold them under observation (or in quarantine if need be) for a sufficient time before mating; the animals will thus be spared a setback due to change of environment near, or even after, mating.

The foundation stock having been obtained, it is well to maintain them under conditions of husbandry similar to those in force in the parent colony. Any alterations to suit local requirements should be made gradually.

There is a wide divergence of opinion regarding the age and weight at which guineapigs should be mated for the first time. Puberty in the female may occur as early as 4 to 5 weeks, when the animal weighs no more than 200 g, but males are not usually fertile until they are 8 to 10 weeks of age and weigh approximately 400 g. In the author's experience, guinea-pigs are best mated first when they are 12 weeks old, the males weighing 500 g. and the females 450 g.

This regime enables the female to give birth to her first litter before she herself is fully grown and there is, therefore, less likelihood of dystocia due to firm fusion of the sacro-iliac joint of the pelvis. Several breeding systems have been described and success has been claimed for each.

The actual system adopted by a breeder may be decided by the type of equipment and housing available, or the purposes for which the animals are required, while others may have a free choice and will be able to consider the pros and cons of each method.

The guinea-pig, in common with some other species of rodents, experiences a postpartum oestrus accompanied by ovulation, and if mating is permitted at this time the interval between litters will be reduced to the minimum and the number of litters born to the animal over a given period of time will be raised.

Breeding may therefore be either by post-partum mating (intensive or continuous method) or by re-mating the sows after the young have been weaned (non-intensive or discontinuous method). The gestation period varies between 59 and 72 days, the average being 63 days.

Non-Intensive Methods

Sows isolated for farrowing

This method is wasteful of space and labour but is the method of choice where accurate breeding records are required, e.g. for inbred strains and for the production of animals to be issued as litter mates. Breeding units (harems) may consist of from 5 to 10 sows with one boar, and 1 ft^2 (0.09 m^2) of floor should be allowed for each animal.

As the sows approach parturition they are isolated individually in a pen, hutch or box of suitable size (approximately 2.5 to 3.0 ft^2, 0.23 to 0.28 m^2 floor space). After the young are weaned, the sows are returned to the same boar for remating. The usual breeding span on this system is two years, but units with exceptional records may be allowed to continue for up to three years.

Apart from its value for the keeping of accurate breeding records, this system is also excellent for disease-control as units are small and all movements are known. Annual output should average 12 youngsters per sow.

Communal farrowing and rearing

This is a common method of guinea-pig breeding. The sows when heavily pregnant are removed from the mating pens and allowed to litter down in groups: the young are reared communally. The youngsters are weaned when they reach 180 g. The empty sows are then restored to the mating pens and the cycle continues.

Opinions vary as to the optimum ratio of boars to sows. It is common practice to mix the sexes in the proportion of one boar to five sows but, in the writer's experience, the best results are obtained when there is only one boar in each pen, the number of sows being anything up to 40. (It is an advantage when more than 20 females are run together to have two or three boars available, each being used in rotation for one week.)

The breeding span under these conditions is approximately two years, during which time each sow should have produced seven litters. Some breeders prefer to replace the boars after 12 months' use.

This system is economical of labour and provided steps are taken to exclude infertile sows production rates of 10 to 12 youngsters per sow per annum may be obtained.

Intensive Methods

Monogamous pairs

Bruce & Parkes (1948) have reported a high degree of post-partum mating in a small colony bred by this method, which, however, is not recommended because of the large number of boars which has to be maintained. Record-keeping is simple and the method may have a limited use for the selection of animals most suitable for intensive breeding in polygamous groups.

Polygamous groups

A selected number of sows is mated to one boar. The sexes remain together forming a static colony which remains intact throughout its useful breeding life. The young are reared communally and are weaned when approximately 180 g in weight.

Such a colony will breed at a satisfactory rate for 24 to 30 months, during which time they will produce 11 litters or approximately 14 to 16 young per sow per annum.

Opinions differ regarding the optimum number of sows for a polygamous group. Working with pens of a constant size, 24? ft' (2.28 m^2), Rowlands (1949) judged that the optimum ratio was 12 sows to 1 boar.

The author, working with pens of varying size and colonies of varying numbers, has concluded that more important than the number of sows to each boar, is the floor space allowed to each sow. It should not be less than 1.7 ft2 (0.16 m^2).

Satisfactory results have been obtained with colonies ranging in size from 4 to 40 sows mated to 1 boar. From the point of view of disease-control, however, the smaller the group the better, and whilst four sows may be the ideal size, it should not be allowed to exceed 20.

Weaning

Whatever the method of breeding, youngsters should be weaned between the 21st and 28th days after birth or when they are judged to have reached 180 g body weight. After sexing, the youngsters are transferred to stock runs to await transference to the laboratories when they attain the weight required.

In the stock runs each weaner should be allowed a minimum floor space of 1/3 ft^2 (0.03 m^2) at 200 g. rising to 0.4 ft^2 (0.037 m^2) at 300 g and to ½ ft^2 (0.046 m^2) at 400 g. Larger animals require more space, reaching 1.0 ft^2 (0.093 m^2) each at 800 g.

Selection of Breedig Stock

Much effort, time and money are wasted through the use of inferior stock for breeding purposes and when a new herd is being established the breeding performance of the foundation stock should be checked by carefully recording the individual breeding performance of a proportion of it.

If output is lower than average, only youngsters from parents with records above average should be kept as breeding replacements. Such a project has been described by Dunkin *et al.* (1930), starting with a group of guinea-pigs of unknown potentiality. When the herd for which the writer is responsible was founded, a small fullyrecorded section was maintained for some 10 years.

The breeding units were composed of six sows and a boar. The heavily pregnant sows were isolated shortly before parturition and returned to their own boar immediately the young were weaned. An individual record was kept for each sow showing the number of young born alive or still-born, the number of young dying during suckling, and the number weaned.

Coincidentally with the weaning of her third litter, each sow was graded on her record either A, B or C according to Table elsewhere in this chapter. It must be stressed, however, that the youngsters of any guinea-pig whose losses, including still-births, exceed a total of one in her first four litters, or a total of two in all litters, were not used for breeding, and the sow was not retained in the unit. Similarly, all which failed to grade because of small litters were likewise discarded.

Table 3.1: Standards for grading of breeding Guinea-pigs.

Litter	A	B	C
	orderNo.	*weaned in*	*excess of:*
3	12	11	9
4	16	14	12
5	20	17	15
6	24	21	18
7	28	25	21
8	32	28	24

Each litter from grade-A mothers was judged as a whole for uniformity of development and, provided they were up to standard, they were used as replacements for this special group. Surplus youngsters from grade-A mothers were passed out to the production units, any further breeding stock required for these latter units being made up with youngsters from grade-B mothers and, exceptionally, from the grade-C mothers.

Now that the colony has been well established, individual grading has ceased. The entire breeding colony is mated in permanent harems consisting of one male and four females which remain undisturbed throughout their breeding life which may extend up to 30 months.

Replacement breeders are selected from those harems which have been mated for 12 months or longer and from which young have been weaned at an average rate of five or better per month, timed from the date of first mating.

In order to avoid the possibility that the colony might split into definite groups by unintentional inbreeding, the colony is arbitrarily divided into eight units and new harems are set up according to the systematic method of breeder-rotation described by Poiley (1960).

Discarding of Breeding Stock

At the end of the breeding life of a group or colony the boars should be removed, the pregnant sows being left to litter down for the last time according to the system in use. Any animals which are non-pregnant when the boar is removed may be detected by failure to diagnose pregnancy by manual palpation 18 to 28 days later.

Discarded animals need not necessarily be destroyed out of hand. They may be useful for certain experiments requiring especially large animals, and large boars are particularly useful for the provision of complement.

Handling and Identification

Guinea-pigs are probably the easiest of all laboratory animals to handle; they seldom bite. They should be lifted by grasping the trunk gently but firmly with one hand and supporting the hind limbs with the other.

Should it be necessary to immobilize the animal, the thorax should be held firmly between the thumb and fingers of one hand whilst the hind legs are held and extended by the other. Additional control may be obtained if the animal, while thus held, is placed on a table or bench.

In breeding colonies the only satisfactory method of identifying guinea-pigs is by tattoo marks in the ear. Tattooing forceps are employed and the size of the individual letters and numbers is 0.5 cm high × 0.25 cm wide.

Three cyphers may be put in each ear. Black ink is used for light-skinned and green ink for dark-skinned animals. Parti-coloured guinea-pigs may be identified by sketch cards showing the outline of their colours but this is more useful in the laboratory than in the breeding house.

Light-coloured animals may be temporarily identified by staining patches of their coat with dyes such as 1 per cent of gentian violet, fuchsin, or crysoidine in 25 per cent of methylated spirit to which 1.5 per cent of carbolic acid has been added. These dyes, however, are but temporary, and if the animals must be identified for lengthy periods the colours will have to be renewed at intervals.

In many instances it will suffice if pens, boxes, or cages are positively identified and the individual pigs identified by one of the foregoing methods or by punch-holes or snips in the ear. This latter system has its limitations as guinea-pigs often exhibit lacerations of the edges of their ears, and unless the marks are sufficiently large they tend to heal producing small fibrous nodules.

The identification of dead guinea-pigs on their way from the animal house to the postmortem room often presents difficulties. An identity label may be securely tied round the neck, but a method which the writer has found useful is to place the body in a polythene or strong paper bag bearing the identification number clearly on the outside.

ACCOMMODATION AND EQUIPMENT

Housing

Guinea-pigs can be bred successfully under a number of widely differing environmental conditions. In temperate regions the best results are obtained indoors when the temperature is maintained within the range 60° to 65°F (15.5 to 18.5° C).

If the temperature is allowed to go above 90°F (32°C) for any length of time, heat prostration is liable to occur, particularly in heavily pregnant sows, and deaths are not uncommon. If the temperature is not kept above 55°F (13°C) youngsters grow poorly.

To maintain optimum temperature conditions in Britain it is necessary to provide some form of artificial heating for roughly eight months of the year, and an automatic air-heating plant is recommended. When heating is dependent upon hot-water radiators, great care must be taken to prevent falling temperatures during the night and at week-ends when boilers are damped down.

A secondary, thermostatically controlled, bank of electrical tubular or convection heaters set to function if the temperature drops below 13°C (55°F) will overcome this hazard. These heaters are also extremely useful during those nights with abnormally low temperatures which occur in spring and autumn whilst the general heating of a building may not be functioning.

In tropical and sub-tropical regions housing is essential and cooled air must be introduced into the houses to lower the room temperature to the optimum. Good natural ventilation (4 to 6 changes per hour) may be adequate but if the concentration of stock is high and a greater rate of air change is necessary, some form of airconditioning will be essential.

Humidity should be kept about 60 per cent. This can usually be achieved in this country but may rise to too high a level in hot sultry weather. Where hot humid conditions prevail for lengthy periods refrigeration can be used to extract the excess moisture.

Windows should be so arranged as to allow adequate light to all parts of the animal rooms, but care should be taken to allow the animals shade from the direct rays of the sun. Where daylight alone is insufficient to provide good lighting, particularly during the winter, then adequate artificial lighting must be provided.

Periods of anoestrus are not uncommon in females during the winter months, but this can be largely overcome by providing additional artificial lighting to extend the daylight to a minimum of 14 hours. Some breeders achieve a measure of success in unheated buildings, provided wooden hutches are used and plenty of hay is given, but there is a marked dropping off in production during the winter months.

This is due in part to a reduction in the number born and in part to an increase in the number of deaths of recently-born youngsters. When the animals are bred entirely out of doors the drop in production is even more marked during the winter months but in the writer's experience the death-rate amongst young stock is not high.

It is doubtful whether guinea-pigs produced under such conditions are likely to be of the quality needed for laboratory use. Outdoor breeding as practised at Onderstepoort and described by Wilson-Jones *et al.* (1953) is successful where climatic conditions are favourable. Out of doors guinea-pigs must be afforded adequate protection against foxes, dogs, cats and rats, all of which will attack both young and breeding stock. For this reason totally enclosed hutches are essential.

It is also necessary to provide proper protection from rain, sun and strong winds. Indoors, a variety of possibilities present themselves and they may conveniently be divided into: (a) pens on the floor; (b) tiered compartments either fixed or portable; and (c) cages, either mobile on trolleys or portable on racking.

While in many instances the choice of a system is largely determined by the type of building available, due consideration should be given to the problems of disease-control. There is something to be said for splitting up a colony into several sub-units in separate rooms; outbreaks of disease can be more easily limited and, if a slaughter policy is necessary, then the number of animals to be killed will be reduced. On the other hand such division may add to labour costs.

All buildings should be of substantial construction, well insulated, and vermin-proof. The floor, ceiling and walls should be impervious and if the guinea-pigs are to be kept on the floor then it should be adequately insulated to preserve heat, e.g. 2-in (5 cm) steel-floatfinished cement over air bricks or polystyrene sheeting 1 in (2.5 cm) thick.

Care should be taken to arrange that any windows facing towards strong sunshine can be screened in hot weather and that ventilation is adequate.

Floor Pens

This well-tried system has many advocates, but it is extremely wasteful of space and therefore of building costs and heating. In a conventional colony infectious disease may spread rapidly. The size and arrangement of pens in any given room will depend upon circumstances. Suitable materials for partitions are aluminium alloy, galvanized sheet iron, or ½-in (1¼-cm) wire netting on frames. These partitions should be removable for ease of cleaning, and any permanent stanchions and wall fixtures should be easily cleaned.

It is practicable to have a system of pens without having permanent stanchions let into the floor of the house. The pens, 3.5 ft (107 cm) wide and 8 ft (244 cm) deep, range down each side of the room leaving a passage 2.5 ft (63 cm) wide. Aluminium-alloy sheet partitions (16 s.w.g.) 16 in (40 cm) high are held in position on walls by wire guides (10 s.w.g.), and speciallyconstructed footed corner posts, hold one dividing panel and two front panels in position. Thus there are no obstructions or holes in the floor, and the room can be speedily converted to other uses if necessary.

The cleaning of these pens is simple. Additional partitions are placed in the passage, the front (passage-way) panel of each pen is lifted, and guinea-pigs from the pen are driven through. The dividing panels are then removed and the whole side can be cleaned in one operation.

After fresh bedding has been distributed and the guinea-pigs have been returned to their pens, the whole operation is repeated on the other side. Finally, the passage is cleaned. Schemes of this type may allow up to nine-tenths of the floor space to be effectively and continuously utilized.

Another scheme for floor pens uses only half the available space each day. The floor of the room drains from the centre towards the walls. Down the centre of the room is a series of permanent or movable partitions, the pens being formed by removable panels which run outwards to the walls.

The pens cleaned the previous day are bedded down and the guinea-pigs are transferred to them through popholes or by raising the dividing panel. The panels between the dirty pens are removed and the whole side cleaned out in one operation. Owing to the frequency of cleaning in this system. animals may be allowed less space but labour costs are increased.

Floor pens can be arranged to fit any floor and are cheap to install and maintain. The labour of cleaning, if the pens are properly laid out, is not great. The quality of the cleaning and the efficiency of disinfection and disinfestation can likewise be high if the walls and floor are cement and the partitions metal.

A further advantage of floor pens is that the animals can be inspected daily with the minimum of effort and in consequence any unthrifty or sick animals can be promptly detected.

Tiered Compartments

These either may be permanent in the form of concrete or wooden shelves, or they may be portable metal units similar in design to chicken brooders. Because of the greater floor area needed for passages, permanent tiered compartments may not be more economical of space than floor pens unless three tiers are in use. (Four tiers are normally unmanageable.)

Portable metal units on the other hand may have four tiers and be very economical of space. The pens on tiered shelving can be adjustable in size and this feature is one of their great advantages. Those of the permanent type are expensive in labour for cleaning and are generally difficult to disinfect efficiently.

The metal portable units, if fitted with wire floors, are easy to clean and can be removed from the room and subjected to thorough chemical or steam sterilization.

Cages

These may either be carried on trolleys, or placed on racking. A cage system is economical of space but is expensive in first cost. Cages can be constructed in a variety of materials and to any convenient size. The cage illustrated. 8, is 30 × 30 in (75 × 75 cm) in area and 10 in (25 cm) high and is made out of a single sheet of 18 s.w.g. aluminium alloy, the corners being welded.

The base of the box is therefore watertight. Provision is made for introducing a panel to divide the cage into two equal parts. In practice the cage holds a harem of four sows and one boar and their unweaned progeny, or when it is partitioned each half is suitable for farrowing one or two sows or can permanently house a monogamous breeding pair.

Alternatively the cage may be used to house a small polygamous mating group of six sows and one boar, or for holding growing stock when each box will hold 12 youngsters from weaning to 350 g, the numbers being reduced as the weight increases—10 at 450 g, 8 at 650 g or 6 fully grown adults.

Cages with wire bottoms, are becoming increasingly popular and many competent breeders claim excellent breeding and rearing results; others, however, have had much trouble with broken legs and deaths by strangulation in newborn young. If losses are to be avoided, guinea-pigs must be brought up to live on wire so that they learn to walk on it without trouble. If they are reared on solid-bottomed cages and transferred to wire-floor cages in later life, it is wise to keep the wire well covered with hay for the first week or two.

The size of the wire is important. Paterson (1962) suggests that a rigid welded floor of 3 × ½ in (7.62 × 1.27 cm) mesh is ordinarily satisfactory. Hoyland (1963) on the other hand suggests ½ × ½ in (127 × 1.27 cm) for stock being reared and 5/8 × 5/8 in (1.6 × 1.6 cm) for breeding animals. Such cages, which are virtually self-cleaning, need only be 8 in (20 cm) deep with a tray 1½ in (3.8 cm) deep to catch the droppings.

Food and Water Utensils

The type of feed utensil will, to some extent, depend upon the type of food. Dry mixtures or wet mashes may be placed in open troughs of either galvanized iron or glazed earthenware, the size and number being adjusted to suit the particular pen or cage.

Open troughs require a thorough washing daily, and at leastt once weekly they should be steamsterilized or soaked in hypochlorite solution. The increasing use of pelleted food has produced a variety of feeding hoppers. The M.R.C. pattern on the left has a solid base and that in the centre a coarse gauze base which prevents hoppers being clogged by meal.

A guard to prevent wastage may be fitted to the latter type of hopper. The type illustrated on the right has a mesh of such a size that guinea-pigs can pull pellets out with their teeth. The few pellets which fall through when the gridded face is scratched or the whole hopper is agitated by the guinea-pig are picked up and consumed at once.

This latter type of hopper is not suitable for wire-floored cages. Hoppers reduce the amount of faecal and urinary contamination of food and are, therefore, preferable to open vessels. Water, which should always be available for all stock, is best given from an inverted or inclined glass bottle with a metal mouthpiece.

Working on the vacuum principle, this system prevents gross contamination but it should be remembered that infection may gain access to the water along with the air bubbles passing into the bottle and carrying with them particles of food and other material from the mouth and lips.

Guinea-pigs soon learn to operate a low-pressure valve-controlled automatic watering system but they are inclined to play with the drinking point, which will make pens and solid-floored cages very damp. Satisfactory automatic systems are difficult to maintain in satisfactory working order in districts where the water is very hard, i.e. where it contains an abundance of dissolved salts of calcium and magnesium.

Bedding

A number of materials are satisfactory, but care must be taken to ensure that they have not been contaminated by excreta from wild rats and mice. Probably the most popular is shavings from seasoned soft wood.

Peat moss is also valuable but is generally more expensive. Satisfactory bedding materials have been made from by-products of flax and from dried maize (corn) cobs. The above materials are both highly deodorant and absorptive. Shavings from green wood are not recommended, neither is cereal straw, whether long or short.

Oat cavings may be used. A sprinkling of sawdust from seasoned wood should be put in the trays under wire floors. A proportion of the hay given will be used as bedding, but this commodity should be considered as a food rather than as a bedding material.

Some people use hay racks to conserve it, but it should be scattered lightly in the pen as the animals delight to burrow under it.

Cleaning of Pens and Cages

Cleanliness is often overstressed, and good results do not automatically follow meticulous attention to cleaning; nevertheless poor results are often due to failure to keep animals in such a way that they are able to maintain a clean appearance.

Opinions differ as to the interval which should elapse between cage-cleanings. Where self-cleaning, wire-bottomed cages are installed, the trays below should be cleaned and the sawdust replenished at least three times weekly. Where bedding is used, cleaning should be done at least once and, preferably, twice weekly.

Floor Pens

The pens should be cleared of animals and all soiled bedding removed. Depending on circumstances, the pens may be hosed down and left to dry after being sprayed or mopped down with a mild disinfectant solution. The pigs are returned once fresh bedding has been put down.

Cages

Ideally a freshly sterilized cage would be available each cleaning-time, but where this is not possible a proportion of the cages should be sterilized at each cleaning so that all are treated in rotation. Failing this, the existing cage should be thoroughly dry-cleaned before fresh bedding is put in.

All scrapers, brushes or other cleaning equipment should be kept in disinfectant solution when not actually in use. It is an advantage if several of each type of tool are available and they are used in rotation, as this gives each a sufficient stay in the disinfectant solution for it to be effectively treated.

DISPOSAL OF REFUSE

This can be a big problem. Refuse should be removed from the animal houses or rooms in bins or barrows as convenient: a chute from an upper floor to a lower collecting point can be a useful arrangement. In urban areas it is best destroyed in a properly constructed incinerator, but care must be taken to avoid creating a nuisance, and smoke and gases may have to be taken up a tall chimney.

Particular care may have to be taken with bedding that has been sterilized and is damp, especially if animal corpses are being disposed of at the same time.

In rural areas alternatives are available. Refuse may be trucked some distance from any buildings, to be burnt in less elaborate incinerators or simply in heaps, or it may be composted and ploughed back into the land as manure.

Bedding from diseased cages and from experimental quarters should be burnt and not composted. All barrows, bins, trailers or other equipment used for the cartage of refuge should be thoroughly washed down and sterilized or disinfected at frequent intervals.

FEEDING

Cavies are vegetarians by nature and they thrive on a wide variety of diets provided care is taken to include the necessary vitamins, particularly C, K and E, and hay to provide a certain unknown but essential factor or factors. A satisfactory diet will consist of a concentrate in either mash or pelleted form, green food in some form, hay and water.

The rearing of guinea-pigs without green food has been attempted but results have been disappointing. A more satisfactory trace-elemented, vitaminized diet (R.G.P.) has been developed and is described below.

Guinea-pigs have been reared on it successfully through 12 generations but it has been necessary to feed long hay as a supplement, otherwise the animals barber each other severely or they consume quantities of their beddingshavings, sawdust or peat-which leads to loss of condition and poor reproduction. The incorporation of powdered hay in the diet was not effective in stopping either the barbering or consumption of the bedding.

Mashes

Probably the simplest mash is two parts of crushed oats mixed with one part of broad bran which is fed dry or slightly damped. An alternative and possibly better mixture is bran and sugarbeet pulp in the proportion of 2:1. The beet is soaked overnight, the surplus water is drained off and the bran is added and mixed in to form a moist crumbly mass.

Only the required amount of this mash should be prepared at one time as it quickly goes sour. (This also applies to the practice of damping mashes to keep down dust.)

These simple diets, however, lack the necessary protein to maintain a high level of reproduction on one hand and regular growth in young stock on the other. Such diets can be greatly improved by the addition of a protein concentrate with added minerals, such as that given in Table elsewhere in this chapter.

Table 3.2: Protein Concentrate.

	Parts by weight		*Parts by weight*
White fish meal ...	2	Dried skim milk	2
Meat and bone meal	3	Finely ground chalk	1
Linseed cake meal	3	Commercial mineralized salt	1

12 parts by weight of the mixture to be added to 100 parts of mash

At Onderstepoort the mash given in Table elsewhere in this chapter, supplemented by green lucerne and lucerne hay, has proved statisfactory.

Table 3.3: Onderstepoort Mash.

	Per cent		*Per cent*
Maize meal	40.0	Lucerne meal	6.0
Crushed oats	10.0	Bone meal	2.0
Wheat meal	25.0	Salt	0.5
Linseed cake mea	10.0	Calcium carbonate	1.0
Meat meal	5.5		

It is common experience that the feeding of mashes, which can be done only in open dishes or troughs, is wasteful because the animals tend to scratch about in them and because the food soon becomes contaminated with urine and faeces. It has also been observed that the animals practise a certain amount of selective feeding, leaving some ingredients relatively untouched.

Pelleted Diets

Although more expensive, pelleted diets are preferred by most laboratories. Among the advantages which may be claimed for them are the following: (1) nutritional intake is uniform and selective feeding is impossible; (2) pellets can be fed in suspended containers or hoppers, thus eliminating waste and minimizing faecal and urinary contamination; and (3) hoppers make it possible to institute a continuous or *ad fib.* feeding system which makes for simplicity and is labour-saving.

The sum total of these advantages may well be that the real cost of a ton of pellets is actually less than that of an equal quantity of mash, without considering the nutritional benefit which must accrue.

In Great Britain most pelleted diets are based on Diet 18 of Bruce & Parkes (1947), which has the formula given in Table elsewhere in this chapter.

Table 3.4: Diet 18.

	Per cent		*Per cent*
Wheatfeed	15	Barley meal ...	20
Grass meal	30	Meat and bone meal	8
Decorticated groundnut meal	15	Salt	1
Linseed cake	10	Chalk	1

The formula would be improved if the decorticated groundnut meal were replaced with an equal quantity of toasted soya-bean meal. This would rule out the possibility of aflatoxin poisoning. Short & Gammage (1959) have described a diet, S.G.I., which they found to be superior to Diet 18 not only for guinea-pigs but also for rabbits. It has the simple formula given in Table elsewhere in this chapter.

Table 3.5.

	Per cent		*Per cent*
White fish meal (67 per cent protein)	10	Finely ground oats	12
Grassmeal (18 per cent protein)	20	Bran (medium)	18
Bran (course)	40		

Hay of good quality is fed to rabbits in addition, while guinea-pigs receive both hay and green food.

The descriptions of the ingredients should be precise, and compounders making pellets for the writer's colonies are required to work to the formula in Table elsewhere in this chapter, with the supplements given in Tables elsewhere in this chapter.

Table 3.6: Diet Q.G.P.

	Cwt	*Per rent*
English fine wheatfeed	6	15
Finely ground oats (thin husks)	5	12.5
Barley meal (plump grains)	16	40.0
Grass (or Alfalfa) meal (min. 18 per cent protein)	6	15.0
Linseed cake meal (8 per cent oil or higher)	4	10.0
English white fish meal (66 per cent albuminvids)	3	7.5
Vitamin premix	20 lb	—
Trace-elemented mineral supplement	60	—

Table 3.7: Vitamin Premix.

Each 10 lb. of the vitamin premix contains

A stabilized	4,000,000 I.U.	K_3 (Menadione)...	10 mg
D_3	1,000,000 I.U.	Calcium Pantothenate	4 g
B_9	8 g	Choline Chloride ...	200 g
B_{12}	12 mg	Butylated Hydroxytoluene	125 g
E	25 g		

Table 3.8: Trace-elemented mineral supplement.

	Per cent		*Per cent*		*Per cent*
CaO	30..0	$CuSO_4$	0.1	$CoSO_4$	0.03
MgO	1.5	NaCl	30..0	Fe_2O_3	1.25
P_3O_5	10.0	KI	0.03	$MnSO_4$	0.2
S	0.2				

Care must be taken during manufacture to ensure that the vitamin premix and the mineral supplement are evenly distributed throughout the final product. The pellets have the average analysis given in Table elsewhere in this chapter.

Table 3.9: Analysis of Pellets.

	Per cent
Oil	3.6 to 4.0
Albuminoids	18.0 to 19.0
Fibre ...	6.0 to 7.0

By being precise in defining the component ingredients one ensures that batches shall be very uniform whether made by the same or by different compounders. This diet requires to be supplemented with greens to provide adequate amounts of vitamin C.

Freshly prepared pellets made with grass meal or lucerne (alfalfa) meal of high quality may contain Sufficient vitamin C for normal growth but unless carefully pelleted they lose vitamin C rapidly during storage. In addition to greens, good-quality hay must be provided.

Diet R.G.P., which was formulated with the object of eliminating green food from the diet, has the same formula except that 1.0 kg of ascorbic acid is added to every ton (1016 kg) of diet, i.e. 0.1 per cent.

If pelleted in the cold state or with the minimum amount of ,team or additional water, the diet will contain between 90.0 and 105.0 mg of ascorbic acid in each 100.0 g of diet, depending upon the amount of natural vitamin C present in the grass or alfalfa meal.

If the pellets are stored under dry conditions there will be little or no loss of ascorbic-acid activity over six or eight weeks. However if, immediately after manufacture, examination shows that the ascorbic acid is 80 mg or less per 100 g of diet, rapid and progressive deterioration of the vitamin C usually occurs even under good storage conditions.

In the United States, commercial firms prepare pelleted guinea-pig diets which are fortified at the time of manufacture with declared amounts of vitamin C. In view of what has been said above regarding diet R.G.P., it would be better if a guaranteed minimum content after manufacture or on delivery were stated.

The amounts which are commonly added are insufficient to maintain pregnant females in good health: Bruce (1950) suggests that their daily requirement is of the order of 20 mg. Other diets are labelled 'complete' and these contain a proportion of hay.

It is not the practice in the United States to publish the formulae of commercial diets, but the ingredients are listed and a chemical analysis is given. It is suggested that the practice in Britain of keeping to a published formula is a better practice.

The writer has observed two instances in which the substitution of one ingredient for another, without affecting the analysis, caused animals to refuse the resultant pellets for several days.

A typical American diet is prepared from ground whole wheat, ground whole oats, alfalfa leaf meal, soya bean meal, soya bean oil, calcium carbonate (from limestone), iodized salt, bone meal, and irradiated yeast, and has an analysis of 18 per cent protein, 3 per cent fat and 13 per cent fibre.

It will be noted that the striking difference between this diet and those originated in the United Kingdom is the omission of protein of animal origin.

Green Food

A growing guinea-pig weighing 250 g requires 3.0 mg and a fully grown resting pig at least 10 mg of vitamin C daily to prevent the occurrence of scurvy, to maintain natural resistance to disease, and also to keep the blood complement at its normal level.

The term 'green food' must not be taken too literally. Certain roots are valuable during the winter months, and sprouted grain is also worthy of consideration. All the members of the cabbage and kale family are suitable, especially the stalks of the latter, which should be split open.

The white hearts of cabbages and savoys on the other hand are often left untouched. Chicory is a useful crop which provides several cuts a year, as do lucerne, and grass leys containing leafy strains of grass. Of the root vegetables, carrots are excellent, swede turnips are inclined to be too hard, and marigolds, although relished, have a relatively low content of vitamin C.

The type of green food being fed should not be suddenly changed but, alternatively, the writer has found that guinea-pigs do well when different types of green food are offered in rotation on successive days. Green food should be fed as fresh as possible, and any spoiled material should be rejected.

If contaminated by soil the greens should be washed before being fed. In frosty weather, greens should be thoroughly thawed out before being fed. If land is available it is sound practice to grow kale, chicory or lucerne, but roots are best purchased as required.

Alternatively, greens may be obtained by contract with a farmer or horticulturist in a rural area, or with a vegetable wholesaler in an urban area. Where guinea-pigs are kept out of doors in grass runs it is necessary to give fresh greens in the winter months.

Guinea-pigs reared on diet R.G.P. without green food do not readily accept greens as part of their diet. Older animals (9 months or more) require three to four days to become accustomed to them, and young animals, particularly weanlings of 200 to 300 g, may eat very little of them for up to a week.

This must be borne in mind when such animals are transferred to a diet of which green food is an essential part. During the transition period water must be constantly available, for otherwise the animals will become dehydrated even in the presence of plenty of green food. On one occasion a batch of guinea-pigs reared on diet R.G.P. and water were transferred to a distant laboratory and placed in individual cages.

Ample food (crushed oats, bran and cabbage) was offered but nothing was eaten and many animals died within 72 hours. Those that were still alive were offered water and hay and most of them recovered completely within 12 to 24 hours.

Hay

The most suitable hay is fine meadow hay of good quality. It should be free from weeds and particularly from thistles, whose spines injure the mucosa of the mouth. Seeds hay, clover hay and lucerne hay are not readily eaten owing to the coarseness of the stalks they contain. Hay should be fed *ad lib.* to all classes of stock.

The exact function of hay has not been clearly defined. Undoubtedly it prevents boredom and it will, of course, add roughage to the diet. Whether it contains certain vital food factors has yet to be determined.

Water

It is unwise to rely on green food as the sole source of water. Considerable privation may be unintentionally caused by misjudging the amount required or even by missing out an occasional cage or pen, whereas water will seldom be omitted, since an empty water bottle readily attracts attention.

It should not be assumed that guinea-pigs kept out of doors in grass runs will obtain sufficient moisture for their needs, particularly in times of drought, and water should always be available.

LABORATORY USE

Provision of Animals

In many laboratories the animals come from an associated breeding unit or from a single (accredited) breeder: in such cases further quarantine is unnecessary. If, however, the animals come from more than one source, no matter how reliable, each batch before being used should be subjected to a period of quarantine of not less than seven days. Guinea-pigs from different sources should not be mixed unless this is unavoidable.

On arrival at the laboratory, the animals should be carefully inspected, and sexing should be checked. If there has been any mixing of the sexes it should be remembered that some of the sows may have been mated, and if this would interfere with the proposed experiment fresh animals should be obtained. (With a little practice pregnancy can be detected by abdominal palpation as early as 14 days, but it is simplest after 21 days.)

All deaths, whether occurring *en route* or after arrival in the laboratory, should be investigated, and if a death is due to an agent likely to give rise to an epizootic it may be advisable to slaughter the survivors or to return them to the vendor. If the animals are kept, the quarantine should continue for a week after the last death.

Animals awaiting use may be kept as sexed colonies in pens or cages similar to those employed for breeding purposes or they may be held in cages specially designed for the purpose. In many laboratories the animals on receipt are placed in their experimental cages, and when required are moved into the experimental rooms together with their cages.

Cages for Experimental Animals

It has not so far been possible to design a cage which is suitable for all types of experiments in which guinea-pigs are employed. For most studies a cage of 3/4-in (1.9-cm) wire mesh and at least 10 in (25 cm) high is suitable. Cages may vary in floor-size from 10 × 10 in (25 25 cm) for a single pig to 24 × 18 in (60 × 45 cm) for 10 small animals Wire mesh (½ in, 1.25 cm) is suitable for flooring, with a tray underneath containing sawdust to catch faeces and urine.

Other workers prefer their animals to sit in the tray on bedding composed of either seasoned wood shavings or peat moss. Cages containing animals are usually stacked in racking or shelving, which should be of a type which can be adjusted to suit the particular cage in use.

In order to facilitate sweeping of floors and cleaning of walls, racking and shelving should be built either as trolleys or as easily portable units, or should be constructed sectionally to permit easy and rapid erection and dismantling.

Where airborne infections from one guinea-pig to another must be reduced to a minimum, cages with solid sides, back and roof are frequently used. Where agents of exceptional risk for human beings are involved, watertight containers with solid walls and a ventilated lid, such as a dustbin, must be used. People handling or cleaning such animals must wear special protective clothing.

Selection of Animals for Experimentation

Each animal must be handled individually and any which are below par in any way should be rejected. The sex should again be checked-a point often overlooked-and if there has been any mixing of the sexes, then the sows should be returned to stock until it has been determined whether or not they are pregnant.

One point to bear in mind is that guinea-pigs take unkindly to being moved from one environment to another, and this is reflected in a marked, but temporary, loss of weight following the move. A 350-g animal may lose as much as 50 g in 24 hours. This loss is regained in the following 48 to 72 hours.

In any experiment in which weights are critical the guinea-pigs should be first placed in their experimental quarters, and observations should not start until the daily weight-gain is normal. On no account should the location of an animal be changed during the course of the observations.

LABORATORY PROCEDURES

Certain nutritional studies, particularly the biological assay of vitamin C, are carried out in guinea-pigs, but the main use of this species is for bacteriological research and diagnosis. This is because guinea-pigs are susceptible to quite a large range of infections from which they do not normally suffer, e.g. tuberculosis, anthrax and leptospirosis.

Feeding

By pipette

Small amounts of liquid (up to 5 ml) may be given by pipette. This is more easily and accurately done by two persons, one handling the animal and the other using the pipette. If the handler exerts firm pressure on either side of the lower jaw the animal will open its mouth and the liquid can be dropped on to the tongue.

By stomach tube

This is the method of choice for larger amounts of liquid or of solids suspended in liquids. The tube, made of rubber or polythene, must be soft and pliable and about 1/16 to ¼ in (1.57 to 6.3 mm) in diameter.

With the animal lightly anaesthetized the end of the tube, which should be rounded, is carefully passed over the tongue into the pharynx, to be swallowed.

It is then gently passed down the oesophagus into the stomach. Care must be taken to ensure that the tube has not passed into the trachea, as may be indicated by a degree of respiratory distress or be detected by listening to the free end of the tube for the passage to and fro of air. When satisfied that the tube is safely in the stomach the operator can inject the agent by syringe.

By admixture with the diet

If the quantity is small, either liquid or solid may be poured over or added to a small portion of the animal's normal diet. Normal diet should be withheld until all the treated food has been consumed.

When large. amounts of an agent are involved, or when the normal diet has to be changed, the change to the test diet should be made over the course of three to four days.

Inoculation

Intradermal

As a preliminary the hair must be removed. From minute areas it may be plucked, but otherwise it should first be clipped and then shaved or treated with a depilatory substance. A fine needle ¾ in (1.9 cm) long will generally be satisfactory.

The skin of the selected area, after sterilization, should be tensed and the needle be inserted to its whole length within the skin substance. The needle should be slowly withdrawn as the injection is made. Up to 0.5 ml may be injected at one site.

As an alternative to inoculation, superficial scarification may be practised. After the skin has been sterilized and the test agent has been placed on the skin, several scratches are made in the area of the skin covered by the agent. Afterwards, any excess agent should be removed.

Subcutaneous

This may be done in a variety of positions as the skin of the guinea-pig is very loose, and the position chosen may depend upon the purpose of the observation. The hair at the site of inoculation should be clipped and the skin treated with an antiseptic.

A tent of skin is raised and the needle (20 s.w.g. × 1½ in, 38 mm) inserted under the skin, parallel with the underlying muscle. The site should be sterilized again after the withdrawal of the needle. This latter precaution is to protect handlers from infection by leakage and also to prevent cross-infection of other animals in the same cage by the same means.

Intraperitoneal

A site on the abdomen, 1 in (2.5 cm) square, slightly to the right of the midline and I in in front of the pubis, should be prepared for inoculation. The guinea-pig should be held on its back with the head held slightly below the level of the hindquarters to allow the stomach and intestines to drop forwards.

The needle (20 s.w.g. × 1½ in, 3.8 cm) is inserted as for a subcutaneous inoculation, but once through the skin the needle is turned at right angles to the skin surface and passed gently through the muscle layer into the peritoneal cavity. Withdrawal of the needle is made in the reverse manner and the site again sterilized.

Intramuscular

The place of choice is the muscles of the thigh. A site is prepared on the posterolateral aspect, about ¼ to ½ in (6.3 to 12.6 mm) above the stifle. The needle is passed through the skin and directed into the muscle mass and the injection is made. Care must be taken not to go too deeply and hit the femur with the point of the needle. A suitable needle to use is 20 s.w.g. × 1½ in (3.8 cm).

Intravenous

Although seldom practised, two methods are possible. The first, which is really practicable only in large animals weighing over 450 g, is to inoculate into the marginal ear vein using a fine, fairly short, needle (23 s.w.g. × 1 in, 2.5 cm).

In order to ensure that the ear veins are readily visible, the operation must be done in a very warm room and it may help if local heat from an electric lamp is available. The guinea-pig must be firmly held by an assistant who can further help to distend the vein, for ease of entry, by pressing lightly at the base of the ear.

The second method is to inoculate into the saphenous vein at the point about half an inch above the hock where it passes subcutaneously inwards and upwards.

The vessel, however, is very mobile, and to ensure entry into the vein it is necessary to anaesthetize the animal and, after preparing the site over the vein, to incise the skin for about 3/8 in (9.4 mm), and locate and fix the vein for inoculation.

After this has been completed, a single suture is used to pull the edges of the incision together. This may be removed after seven days.

Intracranial

The animal must be deeply anaesthetized. The hair over the site of inoculation must be clipped, shaved and sterilized. A short incision is made through the skin to expose the cranium. With sterile precautions, a small hole is carefully drilled with a dental drill through the bone to the dura mater. Inoculation is made into the cavity through the dura mater, using a short, blunted needle (20 s.w.g. × ½ in, 1.25 cm).

Collection of Blood

Small samples of blood, 0.25 to 0.5 ml, may be obtained occasionally from the marginal ear vein by simple venesection. The only satisfactory way of obtaining larger quantities is by heart puncture, and this is not without risk to the animal.

The latter must be anaesthetized. It is placed on its right side and the forelegs are drawn firmly forwards. An area, ½ in (1.25 cm) square, covering the fourth and fifth intercostal spaces and 1 in (2.5 cm) from the midline, is clipped and sterilized.

The position of the heart is located by visual observation and digital palpation; a needle (20 s.w.g. × 1½ in, 3.8 cm) mounted on a syringe is inserted through the intercostal space into the heart.

The blood is steadily withdrawn into the syringe, and when withdrawal has been completed the needle also is carefully withdrawn. Casualties arise mainly from haemorrhage into the pericardial sack through rents in the myocardium, but the risk of this mishap is greatly reduced if careful attention is paid to anaesthesia on one hand and the careful insertion and withdrawal of the needle on the other.

The larger the animal the easier it is to carry out the operation. From an animal 350 or 400 g in weight 10 or 15 ml of blood can be obtained.

ANAESTHESIA AND EUTHANASIA

Anaesthesia

The guinea-pig is seldom used in experiments requiring anaesthesia. For anaesthesia of short duration ether is satisfactory, the animal being fasted for 12 hours beforehand.

The simplest method of induction is to place it in a ventilated glass container with a pad soaked in ether, care being taken that the pad does not touch the animal, as ether burns mucous membranes. The animal can thus be observed and be withdrawn when the desired stage is reached. If it is desirable to prolong the period of anaesthesia, as ether air mixture should be pumped into a close-fitting mask.

The addition of small quantities of chloroform (1 part) to ether (25 parts) has been recommended by Shera (1944), and May (1944) states that ethyl chloride produces a quicker and smoother induction.

Barbiturates are useful as anaesthetics, and prolonged anaesthesia may be induced by pentobarbitone sodium given intraperitoneally. The best dose is 0.2 g per lb (0.44 g per kg) of body weight; twice this dose is usually fatal, and half produces but a transient incoordination.

The time taken for deep anaesthesia to develop varies with the individual animal, but is usually complete in 15 minutes and lasts for one to two hours. The injection of additional amounts of this anaesthetic is not advised; an ether/air mixture may be used to reinforce the anaesthesia if it is not deep enough.

Recovery is slow and may take up to 6 or 12 hours. It is essential to keep the recovering animals very warm during this latter stage.

Euthanasia

Poisoning with chloroform, carbon dioxide or coal gas in a properly designed and ventilated chamber is a practical method. When either carbon dioxide or coal gas is used it should be introduced slowly, causing the animal to become drowsy and to die without struggling.

If the gas is introduced too rapidly there is likely to be a great deal of struggling owing to partial asphyxiation. Probably the commonest, quickest and most humane way to kill guinea-pigs is by preliminary stunning followed by cutting the throat. Stunning is most easily carried out by either:

(a) a smart forward blow behind the head with a blunt instrument, or

(b) striking the back of the head and neck *very hard* against a solid horizontal surface, e.g. a sink or table.

In the absence of any such equipment the guinea-pig's neck may be dislocated as follows: with its back towards the operator and its head held between the forefinger and middle finger so that the head is in the palm of the hand, the guinea-pig's neck may be dislocated by a short, sharp downward movement of the wrist, ending in an equally short, sharp upward movement of the whole forearm. Dislocation of the neck causes extensive haemorrhage in the tissues of the neck, and where the nasopharynx has been punctured there is usually some haemorrhage from the nose.

It should be borne in mind that the shedding of blood may be dangerous if the guineapig is suffering from any infection, and in such cases a lethal chamber should be employed.

Transportation

Within the breeding unit and in the laboratory, guinea-pigs may be conveniently carried in open boxes. These may be made of wood, preferably resin-bonded plywood for lightness, or of aluminium.

For journeys by road, rail, or sea stout wooden boxes or wicker hampers are suitable for repeated use. They should be 10 in (25 cm) deep, but the other dimensions will depend upon the size and number of the animals to be sent.

The boxes need to be handled with care or they rapidly fall to pieces. Hampers are light and, if well protected at the corners with rawhide, have a long life: they should be fitted with a light metal tray, 2 in (5 cm) deep to hold bedding.

Non-returnable cardboard boxes, however, are preferred for the transportation of guinea-pigs between breeding and user units. For air travel, tiered crates are useful. They consist of a light batten framework with base, roof and intermediate floors made of resin-bonded plywood and placed 8 in (20 cm) apart.

Provision is made for a small door in the front, and the sides are then lined with -in (19-mm) wire netting and covered by a layer of strong Kraft paper. The animals should be given enough room to move about freely, more room being given per pig in warm weather than in cold.

In cold weather a plentiful supply of bay should be placed in the box, mainly for warmth, but only a small quantity, mainly for food, should be included to cover the requirements of the animals during a summer journey. If the journey is likely to last longer than 12 hours, carrots are preferable to kale, other greens, or lucerne, which soon become very wilted and unacceptable.

In addition to moisture, carrots will also provide a certain amount of food. If journeys are to last longer than 24 hours provision should be made for proper feeding and watering by the carrier: full instructions for this, and the necessary food, and food and water containers, should accompany each consignment.

Great care should be taken to keep the sexes separate during transportation, for otherwise a proportion of the sows will undoubtedly become pregnant and therefore unsuitable for most laboratory procedures.

DISEASE-CONTROL

One of the most difficult problems in the husbandry of conventional colonies of guineapigs, as of other laboratory animals, is the control of disease. Where animals are recruited from several sources or from a dealer, outbreaks of disease must constantly be reckoned with.

On the other hand laboratories which have closed colonies, or draw their animals from breeders with closed colonies, may enjoy long periods of freedom from diseases, though vigilance is always necessary. Every colony. no matter how healthy, probably carries a variety of bacteria and viruses ready to produce disease if allowed to do so by some alteration of environment, of nutrition, or of husbandry.

An outbreak of disease where such a trigger mechanism is involved may be self-limiting once the fault has been remedied. However, with communicable diseases more drastic measures may be necessary, and these may range from the destruction of all afflicted animals, and their in-contacts, to vaccination or chemotherapeutic treatment of all animals at risk.

Much of the success in diseasecontrol depends upon the quality of the animal quarters, particularly as regards temperaturecontrol, freedom from damp and draughts, and the exclusion of vermin.

The first essential in the control of disease is to establish the cause of any loss. Where facilities exist, post-mortem examinations should be made on all animals dying in the breeding and rearing colonies. Routine bacteriological and parasitological examinations should follow where necessary.

These steps will ensure that none of the serious epidemic diseases will be able to obtain a real hold on the colony before control measures can be put into effect. An adequate record of deaths and their causes should be kept. This will serve to focus attention on deficiencies in management, feeding, and disease-control, and will stimulate a desire to make improvements which will lower the casualty rate.

Within the holding unit of a laboratory, and in the experimental house, the examination of each death is not so important, as so much less is at risk. Nevertheless any rise in the normal death-rate should be carefully investigated by the officer in charge of the animal house.

Frequent consultations between this officer and the head animal technician will help to maintain a high standard of disease-control. The responsibility of the individual research workers to report cases of intercurrent infections amongst their animals should be made clear to those concerned.

The symptoms shown by guinea-pigs affected by the commoner epidemic diseases are not pathognomonic of the disease concerned. Some diseases, e.g. salmonellosis, run a rapid course and the animals die within a few hours of first becoming ill, but most guinea-pig diseases run a more chronic course and the affected guinea-pigs gradually lose weight (they feel light when handled) and their coats become harsh and rough to the touch.

It will be convenient to consider the more common diseases grouped according to their causes.

Dietetic Illness

It is possible that many minor upsets in health are due to sudden changes in the diet, but the growing adoption of pelleted diets has made specific dietary disease something of a rarity. However scurvy, which is due to a deficiency of vitamin C (ascorbic acid), is met with from time to time. Vitamin C is stored by the body, but there is a time limit on its availability.

A guinea-pig placed on a diet entirely devoid of vitamin C will grow normally for a few days, but from the 11th or 14th day onwards there will be a progressive difficulty in walking and steady loss of weight, and death will invariably follow in a further 7 or 12 days.

On post-mortem examination the carcase will be extremely emaciated and there will be a characteristic prominent enlargement of the costochondral junction of the ribs. There will also be petechial haemorrhages on all the serous surfaces.

Scurvy may occur when green food is scarce, particularly during periods of hard weather, and when this happens pregnant sows, whose vitamin-C requirement is very high, should receive special consideration.

The exact function of well-made hay in the nutrition of the guinea-pig has yet to be determined. If young, actively-growing stock are deprived of hay, a high mortality may ensue but can be halted by the reintroduction of hay, and the writer has noted severe and rapid dental overgrowth in suckling sows deprived of hay.

This overgrowth of the teeth first prevented the animals from chewing their food properly and thus caused a rapid loss of condition; later, prehension became difficult and food intake became so small that the animals died from scurvy although green food was freely available.

Guinea-pigs kept on wire floors and deprived of hay indulge in barbering, and it is not unusual to find young suckling animals completely stripped of hair by their mothers or fathers. While other animals' hair is usually attacked, some guinea-pigs, particularly those caged alone, bite their own hair, usually in a symmetrical pattern, leaving untouched only those parts of the head and neck which they cannot reach.

A further disease which the writer has seen on numerous occasions, and which probably has a dietetic origin, is soft-tissue calcification. This condition was principally found in adult breeding boars aged 12 to 18 months but it did occur in females, and was seen in animals as young as 3 months.

Characteristically, deposits of calcium were found in the myocardium, in the stomach wall, and in the kidneys, but the most striking picture was seen at the flexure of the colon, where massive calcification extending distally from this point for 1 or 1½ in (25 or 38 mm) was so severe as to cause a marked degree of bowel occlusion.

As this occlusion developed, so the passage of food through the gut was slowed down. There was a steady loss of bodily condition, and the animal eventually died. The diet fed to affected animals was Parkes's diet 18 with hay and green food, but the incidence was low, less than 1 per cent, of the males used for breeding.

The condition disappeared when Diet Q.G.P. was substituted for Diet 18. It was first thought that the more regular intake of vitamin D and the narrowing of the calcium: phosphorus ratio were effectively preventing the deposition of calcium in the soft tissue; more recent studies, however, have indicated that the condition does not occur when there is a balanced interrelationship between magnesium, calcium and phosphorus and it is more likely that the addition of minerals containing an adequate amount of magnesium was responsible for the improvement.

The presence of unsuspected deleterious substances in the diet has given rise to losses. Wright & Seibold (1958) have reported progressive infertility following the use of feed accidentally contaminated with an oestrogen. The condition was irreversible.

Paget (1954) described an outbreak in a colony of guinea-pigs of exudative hepatitis and this was typical of many similar outbreaks which occurred in Great Britain about the same time and in later years.

The cause remained unresolved until Paterson *et al.* (1962) showed that the causal factor was a fungal toxin (aflatoxin) present in the groundnut-meal fraction of the diet. The sporadic nature of the outbreaks was due to the fact that many consignments of the meal do not contain any aflatoxin.

BACTERIAL DISEASE

Salmonellosis

Salmonellosis is probably the most lethal of all guinea-pig diseases.

Most commonly *Salmonella typhimurium is* the causal organism involved, but S. *enteritidis* or *S. dublin* may sometimes be the cause. Infection is by ingestion.

The disease usually begins with one or two deaths followed shortly after by an explosive epizootic which may result in the almost total destruction of the colony. More rarely a first epizootic wave gives way to a series of more chronic cases, the animals surviving for two or three weeks before dying.

Some may even recover, and some of these may become carriers. These latter are liable to give rise to future outbreaks of the disease when a fresh (non-immune) population is at risk. Occasionally outbreaks may be confined to cases of the chronic type.

In the acute form the only departure from normal seen at post-mortem examination is a slight enlargement of the spleen, although a diffuse enteritis is not uncommon. The invading bacteria can, however, be recovered from almost all parts of the body. In chronic cases, lesions are fairly characteristic.

The spleen is enlarged and contains numerous small necrotic foci which do not project above the surface of the organ. Similar greyish-white foci are also found in the liver. The mesenteric lymph glands may be enlarged. In the chronic form, the causal organism may be more difficult to isolate, but it can usually be found in the liver and in the contents of the gut.

Isolation and identification of the causal organism is necessary to make a diagnosis. If the disease has been diagnosed promptly it can be brought under control by sectional slaughter, i.e. the slaughter of all in-contacts together with those in the immediately surrounding units.

In most instances, however, the disease is not diagnosed until it has obtained a firm foothold, and then the only practical measure is to kill off the affected group of stock and start afresh.

After all stock have been killed and bedding has been disposed of, the internal surfaces of the room or group of rooms which housed the infected animals should be thoroughly cleansed with really hot water containing 5 per cent of washing soda, and similar attention should be given to caging, trays, pen partitions, feeding utensils and the like.

After cleansing, the whole outfit should be sprayed with an approved disinfectant, and left for 24 hours before being washed down with clean water. Alternatively, to achieve the highest possible kill of bacteria and viruses, the rooms should be fumigated with formaldehyde and steam generated by boiling formalin (40 per cent solution of commerical formaldehyde) and water in suitable open vessels to near dryness. For each ft³ of a bare room 0.5 ml of formalin and 1.5 ml of water should be allowed (17.3 and 52 ml per m³).

Where the room contains equipment (cages, racking, etc.) twice these amounts should be used. The steam penetrates to all available surfaces where it condenses; the formaldehyde (gas) then dissolves in the condensation to produce a 5-per-cent to 10-per-cent solution which is not only bactericidal but also sporicidal.

It also has a high activity against all known viruses. It has, however, no activity against coccidia or the eggs of internal and external parasites. All ventilators, windows and doors must be properly sealed during the fumigation process, which should be continued for at least 24 and preferably for 48 hours.

Where it is not possible to carry out the above-mentioned procedure, formalin vapour may be generated by dropping potassium permanganate into formalin (40 per cent) at the rate of I lb of permanganate to I pint of formalin for each 1,000 ft³ (1.6 kg and 2 1 per 100 m³) of room space to be treated.

The formalin is placed in each of the required number of galvanized or enamel buckets, which should be stood on bricks to prevent damage to the floor by heat. Starting with the bucket furthest from the door, the permanganate is dropped quickly into each bucket in turn. Formalin vapour is generated immediately.

When the operator has withdrawn, the exit door is sealed and the fumigation is continued for 36 or 48 hours. In order to enter the building or rooms to restore the ventilation, it will be necessary to wear a protective gas-mask.

Care must be taken, when these methods are used, to avoid leakage of formalin vapour into the building generally, either by seepage or through the ventilation system, particularly where there is recirculation. Special care must also be taken to ensure that windows be opened to allow the formalin to escape at a speed which will prevent annoyance to persons in the neighbourhood.

The attendant's working clothes must he laundered, and the possibility that infection may be lurking in associated rooms and passage-ways, in food and refuse bins, and in drainage channels should not be overlooked.

After disinfection has been completed, the premises should be left empty of guinea-pigs for at least two weeks. A small number of fresh stock should be introduced, and if no cases have occurred after 28 days, additional animals may be imported from the same source to increase numbers to the desired level. After effective formaldehyde fumigation, animals may be reintroduced as soon as the quarters have been properly ventilated.

Several attempts have been made to immunize breeding stock against sahnonellosis. but immunity has not regularly been at a high level, and fresh outbreaks have occurred.

Pasteurella Pseudotuberculosis

Pseudotuberculosis is the name given to a chronic disease which is caused by this organism and is characterized by the development of caseous nodules, particularly in the lymph nodes and viscera. Infection is usually by the mouth, and the earliest lesions are found in the Peyers patches of the small intestine and in the associated mesenteric lymph node.

Occasionally the disease may start in the retropharyngeal or submaxillary nodes. More rarely, primary lesions may be found in the superficial nodes, e.g. prescapular or inguinal lymph glands when infection has entered through wounds caused by fighting, or in the retropharyngeal nodes when infection has been through abrasions of the buccal mucosa or through the tonsil.

The individual lesions begin as small necrotic foci. As these enlarge they bulge beyond the surface of the gland or organ, and when cut they are found to contain smooth, thick, creamy pus.

Three types of the disease are recognized. The first is an acute septicaemic type with rapid death, probably caused by the rupture of a mesenteric-gland lesion into the mesenteric vein or the cisterna chyli. Miliary lesions are usually discernible in the liver and lungs, but occasionally they may be only microscopic in size.

The second type is the well-recognized chronic disease characterized by progressive emaciation and death in three or four weeks. Nodular lesions varying in size from a pinhead to a hazel nut are found in the mesenteric and other abdominal glands and are scattered throughout the substance of the spleen and liver.

Less commonly similar lesions may be found in the lungs and on the pleura and peritoneum. In the female the mammary gland may be affected and sometimes the wall of the uterus may show lesions. The young of such animals may be congenitally affected, or become infected shortly after birth.

The third type of the disease is usually non-fatal, as the lesions are confined to the lymph nodes of the head and neck. Here the danger lies in the risk that these lesions may burst externally and spread infection widely. Some breeders claim that a large measure of control over the disease may be obtained by vacinating the breeding flock with a heat-killed, polyvalent vaccine, but confirmation is awaited.

A more satisfactory method for controlling this disease has been described by Cook (1954). It makes use of the fact that enlarged infected mesenteric and colonic glands can be palpated manually through the abdominal wall of the living animal. In the early stages the gland swells uniformly, but later it becomes knobbly.

While the disease remains confined to the abdominal glands it is unlikely that the animal is infectious to others and thus the detection and elimination of animals with enlarged glands quickly brings the disease under control in groups of weaned stock.

It should be borne in mind that superficial glands may be affected, and any enlarged glands detected during handling should be regarded with extreme suspicion. In breeding units, although palpation of pregnant sows is not a practicable procedure, the affected unit should not be destroyed.

The pregnant animals may be allowed to litter down and only be remated if found to be free from discernible infection. Palpation at weekly intervals is essential when affected groups are being tackled, but it will become sufficient to palpate at fortnightly intervals as the disease is brought under control, and at monthly intervals as a routine measure.

All youngsters should be examined as they are weaned, as this is a most useful measure in the early location of infected breeding females.

Streptococcus Pneumoniae

This infection is characterized by inflammation of the peritoneum, the pleura and the pericardium, which are covered with a thin fibrinous film. There is an abundant yellowish exudate in the associated cavity. Not uncommonly a gangrenous or caseous pneumonia is present as a terminal feature of the disease.

Particularly is this so if an outbreak occurs among breeding sows, when the condition is frequently seen shortly after parturition. The disease may be confined to the genital tract, with primary abscesses in the uterine wall or the broad ligament.

Pneumococci can be isolated from these lesions and typical organisms are invariably present in large numbers in smears. A feature of the disease is that few premonitory symptoms are shown by affected animals before they die.

Control measures are difficult, but as the disease is most troublesome in draughty, damp, ill-ventilated, cold houses, and under conditions of overcrowding, attention to these deficiencies usually results in a cessation of deaths.

Streptococcus Pyogenes

Infection of the lungs results in a lobar pneumonia which may cause severe hepatization. There may also be an associated haemorrhagic pleurisy. The causal organisms belong to Lancefield's Group C. Affected animals lose condition fairly rapidly.

They cough and sneeze and sometimes have a watery discharge from the nose and eyes. With careful observation it will be seen that some are passing red-coloured urine. In addition to changes in the lungs, varying degrees of myocarditis and percarditis are common.

A more chronic type of streptococcal infection also is recognized. It commonly takes the form of a lymphadenitis of the cervical glands in the neck region, when the disease resembles infection with *Streptohacillus moniliformis*. Purulent arthritis and cellulitis also occur, and when pregnant females are affected abortion often ensues.

As with infections of the pneumococcal type, positive control measures are difficult to institute although, this being mainly a disease of young stock, rigorous culling with isolation of affected groups may prove useful in limiting its spread.

Other Respiratory Infections

Inflammation of the upper respiratory tract with or without lung involvement may be due to Friedlander's bacillus, *Bordetella bronchiseptica* or *Pasteurella spp. As* a rule mortality is heavy only when conditions are bad.

Middle Ear Disease (Labyrinthitis)

This is not uncommon and is characterized by the tilting *of* the head to one side. The tilting may become so severe that the animal is unable to walk in a straight line. Ultimately, progressive into-ordination occurs and the animal is invariably culled.

Suppuration in the middle ear is readily detectable. The prompt elimination of all cases as soon as they are noticed rapidly reduces the incidence of the disease, and in a closed colony complete eradication is usually easily achieved.

Lymphadenitis

Apart from pseudotuberculosis and streptococcal infection of the cervical lymph elands, three further types are worthy of mention:

Cervical adenitis due to Streptobacillus moniliformis

This disease has been well described by Smith (1941). Large. often unilateral, abscesses develop in the neck region. They do not appear to be painful, and they have little or no systemic effect unless they are interfering with deglutition.

They may undergo spontaneous regression. Occasionally the lesions break down and rupture externally. The disease may be reproduced by feeding material contaminated with pus from natural lesions. Infection is probably by way of abrasions of the buccal mucosa.

In the author's experience, the occurrence of fresh cases ceased in a colony when a bad sample of hay containing many thistles was replaced by a good hay consisting mainly of soft meadow grasses.

Mucorm ycosis

This benign infection of the mesenteric gland is due to moulds *of* the *Absidia* spp. and is mentioned here because it may be seen at post-mortem examination and be mistaken for pseudotuberculosis, Fig. 23. Young animals (200 to 300 g) are more commonly affected than older stock.

The gland may be enlarged or it may be bosselated. On hemisection, free caseous pus will exude from one or more centres. In this pus the hyphae of the mould may be readily demonstrated in wet smears prepared by mixing the pus with an equal quantity of 10-per-cent potassium hydroxide solution.

The condition is generally quite benign, and the lesions resolve in from 14 to 28 days, but there is a fair amount of residual scar tissue. Occasionally in young animals infection may spread to the spleen and/or kidneys, with rapidly fatal results. Infection is from hay, particularly new-season's hay, even though it has been well made, or from mouldy hay of any age.

This condition can cause confusion when palpation of the mesenteric glands is being used to control an outbreak of pseudotuberculosis. It cannot be distinguished by manual palpation. In order to avoid unnecessary culling the suspected animals may be isolated and, if the glandular lesions recede within from 14 to 21 days, then *Absidia* infection may be presumed, otherwise pseudotuberculosis should be suspected and the animal destroyed.

A second form of the disease is sometimes seen when veiy mouldy hay is fed to growing stock. The fungus does not invade the tissues but multiplies rapidly within the lumen of the small intestine, provoking a mild enteritis.

Diarrhoea, however, is profuse, and the affected guinea-pigs suffer from severe dehydration which may lead to prostration and death. Animal technicians should be warned not to offer to guinea-pigs any hay which shows evidence of mould-growth.

Salmonella limete

This organism may cause death in suckling guinea-pigs, but in growing animals the infection is benign and causes a simple mesenteric adenitis without pus-formation. Once it has been overcome the gland returns to normal in from 21 to 28 days. Recovered animals act as carriers.

Viral disease

Well established and easily recognizable conditions have not been described for guineapigs.

Guinea-pig paralysis was described by Romer (1911). It occurs sporadically and is characterized by a gradually increasing muscular weakness, particularly affecting the hind limbs and hind quarters. Paralysis of the bladder also was noted. Pathologically the disease was a disseminated meningomyeloencephalitis.

The *Salivary-gland virus* of Cole and Kuttner (1926) and Kuttner (1927) is unimportant. It causes no clinical symptoms in naturally infected guinea-pigs, but can cause death when inoculated intracerebrally into young guinea-pigs.

A pneumonia caused by a virus has been described by Lepine *et al.* (1943). The natural disease is benign and difficult to diagnose. However, the inoculation of infective material produces a well-defined train of symptoms.

The animal becomes sick, and on the third day has a high temperature; it loses weight steadily and dies on the ninth day. Similar viruses have been described by Blanc *et al* (1951) and by Ten Broeck and Nelson (1938).

Diseases characterized by progressive marked emaciation have been described by Pappenheimer and Slanetz (1942) and by Pirtle and McKee (1951). Both conditions were thought by the authors to be viral in origin, but this could not be established.

Protozoan Diseases

Coccidiosis

Coccidiosis is of common occurrence. Only one species, *Eimeria caviae,* is concerned and it is host-specific. The disease has been described in detail by Lapage (1940). Affected animals develop a diarrhoea, become listless, and lose their appetite. Diagnosis may be confirmed by the presence of oocysts in the faeces.

Death from uncomplicated coccidiosis is rare, and the occurrence of the disease in animals which are being adequately fed is rare also. Since the oocysts take a minimum time of six days to sporulate and become infective, control based on regular and adequate pasteurization of cages is effective.

Toxoplasmosis

Naturally occurring toxoplasmosis has been described from time to time. The disease is of a chronic nature, and if allowed to live the guinea-pig will become noticeably cachetic. The most prominent finding at post-mortem examination is a marked increase in the size of the spleen. The parasites may be demonstrated in liver or spleen smears stained by Romanowski's method. The disease is not economically of importance.

Parasites

Lice

Two species are commonly found, very often in association, viz. *Grropus ova/ıs,* Burmeister 1838, and *Gliricola porcelli,* Schrank 1781. Infestation with lice is often unsuspected. but attention will be drawn to their presence on the ends of the hairs of dead animals.

These parasites live on skin debris. They do, however, cause the animals to scratch themselves, particularly at the backs of the ears. If the infestation is heavy, individual animals may show some loss of condition and a harsh coat.

Control is difficult. Dusting with insecticides does not appear to be efficient and dipping the affected animals is not a practical proposition. The use of aerial insecticidal vapours has, however, proved effective in eradicating lice within short periods, with two or three weeks' continuous treatment.

Losses Associated with Pregnancy

Pregnancey Toxaemia

Characteristically the affected animal is in a late stage of pregnancy. It becomes lethargic, refuses food, and dies within 24 hours. At post-mortem examination it is usually found to be in good bodily condition; the foetuses—three or more in number—are not decomposed.

Only the liver is abnormal, exhibiting a yellow or yellow-bronze colour. The disease is probably nutritional in origin and may be due to the intake of food failing in quantity or quality or both towards the end of pregnancy.

Dystocia

Death from inability to deliver the young is not uncommon. Incorrect presentation of the foetus itself does occur, but a large single foetus is a more common cause. Losses are most common in primiparous sows.

Undoubtedly many cases are caused by waiting until females are six months old before they are mated for the first time. This is an unwise practice, for when the sow litters for the first time at the age of 8 to 9 months, her pelvic girdle and ligaments will have become more rigid and be unable to relax to allow passage of the young.

Torsion of the Uterus

This is characterized by sudden death and may occur at any time after the 30th day of pregnancy. The volvulus may consist of part of one horn or a whole horn, or the entire uterus may be twisted. The condition is readily recognized at post-mortem examination, when the affected portions are seen to be severely congested and dark red in colour. The primary cause is not known.

Intra-Uterine Haemorrhage

Deaths from this definable condition have occurred at irregular intervals in some colonies. Only animals past the mid-term of pregnancy were affected. They were rarely seen to be ill: generally they were found dead, and in many cases the fur around the vulva was blood-stained.

In some animals of a pink-eyed white strain seen to be ill, the pupil of the eye, normally a fiery pink, was observed to have become very pale. At post-mortem examination the most striking feature was the massive haemorrhage into the uterus: all organs were paler than normal.

The edges of the liver were rounded and its lobulation was pronounced. The bile duct was prominent and often ballooned along its length. Close examination showed that the common bile duct had been twisted by a partial torsion of the stomach in an upwards and forward direction.

Neil (unpublished observation) has shown that severe partial occlusion of the bile duct, similar to that observed in natural causes, will cause an alteration of liver function whereby one of the factors vital to the normal blood-clotting mechanism (viz. factor VII) is not produced.

It seems likely that small haemorrhages occur in the vascular placenta throughout pregnancy, and that these are normally stopped by coagulation. In a sow low in factor VII the haemorrhage will continue and will terminate fatally.

The underlying cause of the condition lies in the failure to provide palatable roughage in sufficient quantity. Cases of the disease may be precipitated by offering hay of poor quality, when guinea-pigs will consume more pellets.

This causes a change in the nature of the stomach contents, the hay and pellet melange being replaced by a soft mush of pelleted diet only. A stomach with such contents may more easily be pushed upwards and forward by the uterus.

On the other hand, it may well be that the increased protein intake causes more rapid foetal development and that the additional mass of the uterus causes the displacement. Whatever the reason, cases of the disease soon occur when poor hay is fed, and cease to do so when adequate palatable roughage is offered.

Post-Parturient Jaundice

This fatal condition occurs three to seven days after parturition. The animal is usually found dead. The most striking feature is the marked jaundice of all tissues. The uterus is undergoing normal involution.

The stomach, however, has turned over completely and occupies the central abdominal area, with the small intestine filling the space normally occupied by the stomach.

The rearrangement of the abdominal contents has caused a complete twist in the bile duct, stopping all flow of bile to the intestine. Contributory causes of death are occlusion of the pylorus and gastric circulatory strangulation.

Undoubtedly this condition is due to the failure of the stomach to fall into its normal position after parturition. There seems to be no logical explanation.

Still Births

Losses due to still births may be due to individually large foetuses—usually only two or at most, three in number-resulting in prolonged or difficult labour. In a relatively small selected group Rowlands (1949) reported a still-birth incidence of 5 per cent.

In another colony Davis and Williamson (1959) reported a still-birth incidence of 12 per cent, which later fell to 5 per cent. In the Allington Farm colony, when sows are isolated at parturition the still-birth rate is under 2 per cent but in polygamous groups the rate rises to 7 per cent.

Tumours

Spontaneous tumours are rare. Probably the most common are dermoid cysts of the cornea.

Malformations

Davey (1965) has seen one double-headed offspring in 100,000 born. The author has seen a pair of Siamese twins joined at the head, and one youngster with the lower jaw completely missing, in approximately 300,000 young. None of these malformed animals survived.

Mutants

Skinner (1962) has described sticky-haired guinea-pigs in the stock bred at the Department of Pathology, Cambridge. The condition was controlled by a single recessive factor (st) with full penetrance and with no effect on viability or fertility.

There was an excessive amount of lipid which produced an unkempt appearance in affected animals. During one period of three months 5 per cent of the young born were affected.

The writer has observed a single female mutant with a rexoid or fuzzy coat in one litter of the Dunkin-Hartley strain.

Prevention of Disease

In some laboratories the purchase of guinea-pigs is normal practice, whereas in others it is a rare occurrence carried out to meet a demand higher than can be met from homebred stocks. Between these two extremes a very wide variety of circumstances may arise under which purchased animals are introduced into laboratories.

Very special precautions must be taken by laboratories with breeding colonies when they purchase animals for experimentation. In no circumstances must imported animals be allowed to come into contact with the breeding colony.

Great care must also be taken that there shall be no indirect contact by way of things such as carrying-boxes, cleaning-equipment, cages or the like, or that animal technicians carry infection from the imported stock to the breeding colony.

Ideally the breeders will be housed in a separate building, and any imported animals will go straight into the experimental animals' house or into the rooms set apart for this purpose. It is not a good policy to mix home-produced animals with imported ones during the course of experiments, and it must never be done without thought.

If the imported animals have been obtained directly from another closed colony with an excellent record, the risk may be taken after a minimum period of quarantine (seven days), but otherwise quarantine of the imported animals should not be less than 14 days. On the other hand if the stocks are not to be mixed such imported animals may be used, if desired, as soon as they are received.

Where all animals are imported, but from a single reliable (accredited) source, it is unnecessary to subject individual batches to a period of quarantine and the animals may therefore be used as though they were home-bred.

If, however, they have been transported over any considerable distance, a day or two should elapse between arrival and use to allow them to become accustomed to their new environment. Where animals are imported from several breeders or from a dealer or dealers, then each batch received must be held separately in quarantine for at least 14 days before use.

Each batch in quarantine must be inspected daily, and every ailing animal must he culled for post-mortem examination so that, should infectious or contagious disease be disclosed, appropriate action may be taken to limit losses. Quarantine should be continued for at least 7 days after the last death.

4

Locusts

One Result of the increased use of modern insecticides and more efficient methods of warehousing has been greatly to decrease the numbers of cockroaches available to zoological laboratories, where in the past considerable numbers have been used for dissection and as a live food for laboratory animals.

Since Albrecht (1953) described the detailed anatomy of the Migratory Locust and Hunter-Jones (1956) gave instructions for breeding and rearing it, the cockroach has been largely superseded. The locust, being easy to breed and rear has in many ways proved to be more suitable for teaching purposes, and it is also an ideal form of live food for many laboratory and zoo mammals, birds and reptiles.

SPECIES

The African Migratory Locust

(Locusta migratoria migratorioides)

This species is by far the easiest to rear in the laboratory. The young locusts, or hoppers, are about 8.5 mm long when newly hatched, and by the end of the first day they are usually all actively feeding.

Apart from the skin which the newly hatched insect leaves behind as it arrives at the surface of the sand, there are normally five moults occurring at intervals of approximately 5, 4, 4, 5, and 8 days respectively, thus giving an approximate total nymphal life of nearly one month.

The wings are fully formed after the last moult, but the new adults are very soft and the insects will not be able to take flight until the cuticle has hardened. The adults mature in about four weeks after fledging.

When young they are pale grey marked with darker grey, changing to bright yellow in the males and dark brown in the females at sexual maturity. The average female produces six egg pods in her lifetime (about one every 5 or 6 days). There are between 30 and 100 eggs per pod. The incubation period of the eggs is about 16 days at 28°C (82°F), 11 days at 32°C (90°F), or 9 days at a higher temperature, but this is not recommended.

The Desert Locust
(Schistocerca gregaria)

Although somewhat larger in size than *Locusta* this species is not so useful for teaching purposes as it has a longer life cycle and is prone to suffer from a bacterial disease. The freshly hatched hoppers of *Schistocerca* are approximately 9.5 mm long and they usually moult five times, taking 4 to 5 weeks to become adults at the indicated temperatures.

Their colour is controlled to a large extent by the culturing temperature; at 28°C (82°F) the average fifth-stage hoppers are almost wholly black, whereas at 40°C (104°F) they are yellow with faint dark markings. Under the temperature conditions advocated for *Locusta,* i.e. 28°C (82°F) during the night and 34°C (93°F) during the day, the amount of black and yellow pigmentation is about the same.

The young adults are pink but as they age the pink changes to creamy brown which at maturity yellows in the female and turns to bright yellow in the male. Given the recommended conditions and a population-density of approximately 40 pairs per 60 cm^3 the females should lay their first eggs 4 to 6 weeks after fledging, and should produce approximately one pod per week for about 7 weeks.

The number of eggs per pod varies between 40 to 70. The incubation period is 17 days at 28°C (82°F), 12 days at 32°C (90°F) and about 10 days at 34°C (93°F).

The Red Locust
(Nomadacris septemfasciata)

This species occurs in plague form over a great part of Africa south of the Equator. Normally it has only one generation a year, but three generations can be produced in the laboratory. It requires similar conditions to *Schistocerca* and the adults are of about the same size.

The newly-hatched hoppers are about 9.5 mm long and they undergo six or seven moults as opposed to the five of *Schistocerca* and *Locusta.* Some adults may have a very long maturation period, while others emerging at a different period of the year may only require about 6 weeks.

The females lay from four to six pods at an average rate of one a week. The number of eggs per pod is usually about 100. At 32°C (90°F) the incubation period is approximately 22 days.

Egyptian Grasshopper

(Anacridiun aegyptium)

The Egyptian Grasshopper, sometimes known as the Tree Locust, is common in the countries around the Mediterranean. Unlike the other species it does not, however, form large swarms. It is often imported accidentally into England with cargoes of vegetables. The newly hatched hoppers are pale green in colour and about 8 mm long.

They moult five or six times and become adults in from 7 to 8 weeks. They are similar to the previous species mentioned in that some require a lengthy maturation period. The first egg pod is laid from 1 to 6 months after fledging.

Up to ten egg pods can be laid, each containing on an average 50 eggs. The pods are laid singly at intervals of 1 or 2 weeks. At 32°C (90°F) the incubation period is approximately 22 days.

Identifying the Species

If the ventral surface of the thorax is pubescent and there is no peg-like projection between the front pair of legs the insect is the African Migratory Locust. If a peg-like projection is present between the front legs, and the hind wings show a blackish or smoky band when extended this is the Egyptian Grasshopper.

If a peg-like projection is present but there is no smoky band on the hind wing, examine the shape of the projection. It is curved and pointed in the Red Locust, but straight and blunt in the Desert Locust.

Phase and Colour Variation

Locusts are known to exhibit variations in both coloration and behaviour under different environmental conditions. This phase-variation has been shown to be associated with the appearance of the well-known locust plagues in many countries.

Locusts of the plague phase, i.e. phase *gregaria,* are generally very brightly coloured and marked with a deep black pattern. They are also extremely active, excitable, gregarious and continually on the move. In the other phase, *solitaria,* the locusts are fairly inconspicuous in appearance, inactive, and either green overall or brown with no black markings.

Locusta shows the greatest variation, and an initial colony of 1,000 hoppers in a cage measuring 30 to 60 cm^3 will give rise to typical *gregaria* coloration. The hoppers will be glossy black and orange and the young adults will be creamy-white with dark grey or black markings.

Males and females will be of approximately the same size. If, on the other hand, the young hoppers are kept strictly isolated throughout their lives they will show *solitaria* coloration.

ACCOMMODATION

Breeding-cages

Equipment need not be expensive or elaborate, but sound well-fitting cages are necessary as the young hoppers will easily escape through any hole much more than 2 mm in diameter.

It is possible to breed locusts in a cage of almost any size, ranging from about 30 cm^3 upwards, the basic necessities being adequate light and heat and not less than 10 cm of coarse moist sand at the bottom of the cage for oviposition.

Many years of experience have shown that the most suitable type of cage for largescale breeding is undoubtedly one based upon the principles described by Hunter-Jones (1961). Where, however, unheated rooms are to be used, two additional electric lamps can be fitted below the false floor to give the required heat.

The breeding-cage, is in the form of a box measuring 46 × 40 × 24 cm. It is made with a metal angle frame supporting perforated zinc sides and back. Access for working is provided by a hole 11 cm in diameter, cut in the flat metal front and covered with a circular flap.

Observation is made through two glass windows, one on the top of the cage being removable and the other positioned in the front immediately above the covered floor. Three holes, each 3.8 cm in diameter, are drilled at the front of the false floor to accommodate the standard oviposition tubes.

For convenience of working, the sand for oviposition is contained in tubes made of either metal or plastic, closed at one end, and measuring not less than 30 mm in diameter and 100 mm in height.

In order to accommodate these oviposition tubes in the bottom of the breeding-cage, it is necessary to provide a false floor made of either 16-gauge perforated zinc or even hardboard. The distance of the false floor from the main floor is governed by the length of the oviposition tubes.

Normally three or four such tubes should be adequate for a breeding-cage of the size recommended. The holes should be cut in the false floor at the front of the cage, and should be just sufficiently large to accommodate the open tops of the tubes, so that when filled with sand and inserted in position they are flush with the upper surface of the false floor. A 25-W electric lamp is fixed at the rear of the top of the cage for illumination.

Rearing-cages

The rearing cage, is basically a large wooden box 60 cm high by 60 cm wide by 50 cm deep, with a completely detachable lid. A small glass window 20 by 20 cm is inserted in the lid for observation purposes, and there should be a hinged trap of the same size for introducing food, catching up hoppers, etc.

The front of the cage is provided with a large fixed window on one side; on the other side of the front is an adjustable ventilator. Illumination is provided by an electric lamp fitted at the top of the cage. If additional heating is required, extra lamps can easily be fitted.

It is very important to allow the growing insects to perch if they are to moult properly, and so wire netting or dry branched twigs should be placed inside the cage. Lack of adequate perching material leads to crippled insects and unnecessary deaths.

HUSBANDRY

Egg-laying

In their natural habitat locusts deposit their egg pods in the soil. In the four common species these pods are rod-shaped and measure approximately 100 × 8.5 mm. The lower part of the pod contains 30 to 100 eggs, cemented together with a frothy substance which solidifies shortly after being laid.

This substance extends to the surface of the soil and forms the upper half of the pod. In the laboratory, locusts will lay eggs in containers of well-packed coarse moist sand. *Sharp builders*' sand is suitable, providing it is washed to remove dust and then sterilized by heat.

The sand should be mixed with 15 per cent of water and packed firmly into the oviposition tubes, which are inserted in position in the base of the breeding-cage. The number of pairs of adults required per cage is not critical but 30 to 40 pairs are recommended.

The sand tubes should be removed every one or two days and replaced by freshly packed ones, whether egg pods have been deposited or not. It is easy to determine the presence of any pods by gently tapping the top of the tube and pouring off the top surface of the sand.

The frothy plugs at the tops of the pods will then be exposed to view. The tube can then be covered with a well-fitting metal cap. The tubes are put into a 1-kg screw-topped jar, which is dated and put into an incubator set at 32°C (90°F).

An incubator is not essential and any warm place will do. The number of pods laid per tube governs to some extent the number of hatchings expected per pod. If more than one or two pods are laid in a tube then the number of hatchings per pod will be reduced. Six or more pods will probably result in no hatchings at all.

If too many pods are laid in each tube then the number of locusts should be reduced or the number of tubes increased. At from 28 to 32°C (82 to 90°F) the eggs take from 11 to 25 days to hatch, depending on the species.

After the eggs have been placed in the incubator they should not be disturbed until just before their expected hatching. The uncovered tubes can then be put into a new empty rearing cage, so that when the youngsters hatch they will be free to move about and feed; or they can be left in the incubator to hatch in the jars and be dealt with subsequently.

Temperature and Humidity

For breeding and rearing most species the temperature should range from approximately 28°C (82°F) at night to 34°C (93°F) during the day, the length of day being at least seven hours. The temperature of the cage should be read with the bulb of the thermometer about 10 cm above the false floor and about 10 cm from the sides of the cage.

For a cage equipped with electric-lamp heating and standing in a room at normal temperature this would probably require two 25-W bulbs below the false floor and one 60-W bulb in the top of the cage.

Little humidity-control is required. An average of R.H. 45 has given extremely good results, the grass fed during the day providing all the humidity necessary. A very high humidity should be avoided as it tends to encourage the growth of moulds and other pathogens.

When the humidity is unduly high the faeces will be wet and sloppy and condensation will appear inside the cage. This condition should be immediately corrected. Excess humidity is caused by having too much grass in the cage.

Nutrition

For breeding and rearing, most locust species only need a daily supply of fresh grass and dry wheat bran. The grass should be cut into longish lengths and made up into bunches which are held in water-filled pots inside the cage. The actual daily requirement has to be learned by experience. If insufficient is given the locusts tend to cannibalize, and if there is too much the cage will need to be cleaned more often.

Hunter-Jones (1961) recommends that a shallow tray filled with water should be given to adults, but this is not necessary if the grass is fed in water-filled pots. The bran is fed in shallow trays which must not be allowed to become dirty or mouldy.

Normally, fresh grass should be given every day, but it can be omitted on one day a week, provided that bran is always obtainable. It is sometimes difficult to obtain green grass at certain times of the year. If supplies fail an artificial diet may be used, consisting of a dry mixture of bran, dried milk, dried grass and dried yeast in the ratio by weight of 1:1:1:0-1.

Dried chopped grass can be obtained commercially. With this diet the locusts must have access to water. As an alternative, some establishments use wheat grown hydroponically, which the locusts will accept as a substitute.

Hygiene

It is of the utmost importance that scrupulous cleanliness be observed at all times. All traces of uneaten food, dead locusts, and faeces should be removed daily. At monthly intervals, or when the cages become empty, they should be thoroughly scrubbed out with hot water and an efficient detergent disinfectant, followed by a thorough rinsing in clean cold water.

DISEASES

From time to time, locusts cultured in the laboratory are troubled with a bacterial disease. Whereas *Locusta is* rarely affected, and for this reason is the most useful species to work with, *Schistocerca* and *Nomadacris* are much more susceptible.

The symptoms are loss of appetite and an increase in mortality, particularly among hoppers in the fifth instar and young adults which have recently moulted. The body becomes bright pink and its contents almost fluid. The insects have an offensive odour and tend to break up on being touched.

It has been suggested that superficial injuries may assist the onset of this disease, and for this reason Hunter-Jones (1961) recommends that locusts should not be handled when moulting and that there should be plenty of sticks and moulting perches in the cages in order to prevent unnecessary injuries at this time.

The disease is unlikely to occur if the cultures are kept at the correct temperature and humidity, are properly fed, and are kept scrupulously clean. It is possible that the bacteria are always present and only become really troublesome when the locusts are in poor health owing to some other cause. Any infected cages should be cleared and sterilized.

Occasionally the eggs of the parasitic nematode *Mermis* are picked up on the grass used as food. After being eaten by the locust, the young nematode hatches and starts to burrow through the body of the insect until it reaches the body cavity. There it may grow to a length of several centimetres. The remedy is to change the source of the grass used as food.

5

THE MOUSE

MARSUPIAL MICE

Some small members of the marsupial family Dasyuridae are promising laboratory animals. Although predators in nature, they can be kept and bred without live food. The following information is derived from a colony of *Sminthopsis larapinta* (adult weight 20 to 30 g) maintained by the writer and Dr G. K. Godfrey. Notes on keeping another species, *S. crassicaudata,* in captivity are provided by Martin (1965).

Sminthopsis is one of the least difficult of mammals to keep in captivity. A useful feature is its ability to live without food for up to four days. It should be noted, however, that members of the genus are totally protected in all Australian States and may not be collected without a permit.

ACCOMMODATION

As the animals do not gnaw, light wooden or hardboard boxes with fly-wire gauze can be used. A nest-box should be provided for shelter and to facilitate handling. The cage floor is covered with dry sand. Paper strips *(not* cotton wool) are provided for nesting. The animals are kept in pairs or trios.

FOOD

A staple diet consists of minced ox-heart muscle, with chopped hard-boiled egg or mealworms once weekly. Water is always available. Undiluted cows' milk is given to lactating females and newly weaned young. Milk containing a multi-vitamin concentrate is used to moisten the egg, and calcium phosphate is added to the minced meat. Canned lambs' brains (baby food) can also be used.

REPRODUCTION

The females become sexually mature at about 135 days and the males mature later. Gestation lasts 14 days. The litter-size ranges from four to eight. Pouch life is approximately 40 days, but the young suckle for a further 20 days. The duration of the oestrous cycle is at present undetermined. The breeding potential in captivity appears to be three or four litters in a year.

THE DEER MOUSE

The Genus Permoyscus includes over fifty species and more than two hundred described subspecies. It occurs over much of the North American continent. The most common in numbers, the most diversified subspecifically, with over sixty subspecies, and the most commonly kept as a laboratory animal is the species *Peromyscus* maniculatus.

The account of laboratory care and management given here differs in many important details from that in the previous edition of this *Handbook* and a brief explanation of how these differences have arisen is appropriate.

When the present author moved to the Ohio State University from the University of Michigan, after seven years' experience under the management system described by Dr Dice, he was confronted with an entirely different set of physical facilities.

These facilities and the management system described herein turned out to be as successful as those described by Dr Dice. A comparison by the reader of the considerably different systems will show how adaptable the deer mice are.

ACCOMMODATION

The deer mice, a colony of some 2,500 individuals, were housed in the same rooms as various inbred strains of laboratory mice. The two types were handled by similar management methods. The original cages were made of 24-gauge galvanized metal and 4-mesh hardware cloth (copper gauze).

The bottom consisted of a pan 12 × 5¾ in (30 × 14.5 cm) in area and 5¼ in (13.2 cm) in height. The sides of the cages sloped inward slightly, so that the width at the top was 5¼ in (13.2 cm). The bottom and sides were made of one piece.

The front and back were cut from separate pieces which were spot-welded to a ½-in (12.5-mm) overlap folded from the bottom and sides. The bottom seam was also soldered inside, to prevent leaking. The top consisted of a piece of 4-mesh galvanized hardware cloth reinforced along part of the sides and the back by a 1-in (2.5-cm) folded strip of galvanized sheet metal.

The front end of the top was recessed, and the sides at this portion consisted of triangular sections of galvanized metal. The recessed portion was separated into two compartments by another triangular sheet-metal divider. One compartment was used for food; the other provided for a half-pint (1-¼) square water-bottle of the type formerly widely used by commercial dairies.

A one-hole rubber stopper fitted with a stainless-steel tube which was constricted somewhat at the delivery end provided delivery of the water. The portion of the top against which the stopper rested was reinforced with a piece of sheet metal of approximately the same size as the stopper. When in position, the bottle was tilted at an angle of roughly 45° from the vertical.

The front part of the top consisted of a folded strip of galvanized metal which, when in place, hung over the front of the cage proper and provided for a 1 × 3-in (25 × 75-mm) label in addition to supporting the top. The back of the top was supported by another overhanging strip. No support was provided along the sides.

Underneath the front piece of the top a heavy-gauge copper wire was soldered. This wire projected about 1/8 in (3.1 mm) beyond the front piece to either side. Thus the top of the cage could be first lifted and then slid back to any desired extent, whereupon it would be supported in front by the copper wire.

The inward taper of the sides allowed one cage to be removed from the rack without having the copper wire from one cage catch on an adjacent cage. The useful life of this cage is estimated to be approximately three years under the management system used.

Many deer mice have a strong tendency to select one corner of a cage for both defecation and urination. with the result that the chosen corner contains a moist, strongly corrosive pile of faecal matter. Once this has eroded the galvanized coating the underlying steel is attacked rapidly.

A number of methods were tried to prevent or slow down this corrosion, these included the use of different types of rust-proof paints, with and without different kinds of primers, and with and without baking. In no instance did a treatment suffice to provide complete protection, although an application of a polyvinyl resin is now giving promising results.

All cages were housed on commercial steel racks with 30 or 36-

in (75 or 90-cm) shelves, accommodating five or six cages and having intershelf spacing adjustable in 1-in (2.5-cm) increments. For the standard cage, a 9-in (225-mm) spacing between shelves was used.

Each room had a louvered area in the door, and a small blower which exhausted air to the outside. Hot-water radiators provided heat during the winter. Unfortunately no means of limiting the maximum temperature during the surnmer was available and as a result, and in particular because of the southern exposure of a brick-walled exterior, summer temperatures were usually above 80‘F (27°C) and sometimes the room-temperatures rose to 95°F (35 C) with a drop of only three or four degrees F by the following morning.

With these temperatures, which were usually combined with a high relative humidity, there was a noticeable increase in mortality, both adult and juvenile, combined with a cessation of breeding by many pairs.

NUTRITION

Although the individuals which formed the foundation of the breeding colony had been fed with a mixture of rolled oats, grains and supplements the deer mice in the colony were transferred to a commercial ration (Purina Mouse Breeder Chow).

In all cases the transfers were made gradually, over a period of two to four weeks, by supplying mixed diets. As a result of both qualitative and quantitative observations the following general conclusions were reached. Upon change of diets, breeding pairs experienced, for a period of several weeks, a considerably reduced fecundity as compared with simultaneously maintained controls.

Most pairs returned to essentially the original fecundity level within a matter of two or three months. Feeding was accomplished twice a week and consisted simply in filling those hoppers which were less than half full. The hopper would supply a cage containing four mice with enough food for at least one week.

GENERAL MANAGEMENT

The bedding consisted of red cedar shavings, a layer to a depth of roughly ½ in (12.5 mm) was used. No additional nesting-material was supplied, and no special shelter was provided. Cages were changed once each week.

A clean cage with fresh cedar shavings was placed on a work bench and the cage to be changed was set alongside the empty cage. The top, containing the water-bottle, food, and cage label was then moved to the new cage. The mice were next transferred one by one. This was accomplished by catching each one in the hand and putting it into the new cage. *Peromvscus* are considerably more active than the laboratory mice and are rather difficult to catch by the tail.

With a little practice the technique of gently grasping a mouse in the hand becomes easy. Seldom have these mice attempted to bite when handled in this manner. By way of contrast, if rubber-tipped forceps are used to catch the mice by the base of the tail they will turn and attempt to bite both the forceps and any hand which comes close.

It is desirable, particularly for persons who are first learning to handle deer mice, to have the hack and sides of the work bench equipped with a smooth shield, for example 4I -in plywood, to a height of about 1 ft (30 cm).

If a mouse should jump out of a cage during handling it will almost always run away from the worker, when it can be cornered against one of the sides or the back. Cages were washed in a small sink which held two cages at a time. After a set of cages had been emptied of the soiled bedding, a process aided by use of a 3-in (75-mm) putty knife as a scraper, they were immersed in the sink.

One handful of a commercial detergent was used in the water. Each cage was then brushed on the inside with a swivel-joint brush, emptied of water, and stacked upside-down to drain. Cage tops were changed less frequently, or as it was observed that an individual top was becoming dirty.

Tops were washed in the sink used for washing cages. When possible they were allowed to soak for an hour or more, after which they were brushed lightly, rinsed, and set aside to dry.

The half-pint (4-1) bottles used would last a week. The watering system which appeared most satisfactory involved a complete change of bottles in a room (some 500 cages) on one day a week, followed by a spot check four or five days later.

For cages which did not contain more than four mice. the water would last for two weeks. However, the bottles had a tendency to drip when they reached a low level, and the remaining water often became dirty by the end of one week.

BREEDING

All deer-mouse colonies known to the writer are maintained by a system designed to approximate to random mating. A number of attempts to produce inbred strains have failed. Twice in the Ohio State colony, brother and sister matings were carried out until the fifth generation was reached, but the proportion of fertile matings and the average littersize dropped rapidly from the third generation onwards.

It is the belief of a number of persons who have managed *Pcromr'scus* colonies that to pair mice at weaning-time results in a decidedly decreased fertility: no large-scale experimentation has, to the writer's knowledge, been conducted to establish this as a fact.

Accordingly it is the usual custom to segregate offspring by sexes at weaning-time. The normal complement of mice kept in a cage is four. Since the most common litter sizes at weaning are three and four, this means that the progeny must be marked for identification unless they are to be kept one or two in a cage.

The marking method used consisted of a number code punched into the ears with a chick toe-punch. A simple 1 × 3-in (25 × 75-nim) cage label was used to record the strain symbol, the sex, and the individual mouse numbers.

This tag was easily distinguishable from the label used on cages containing mated pairs. Mice were paired at various ages, depending upon a variety of factors. The optimum age for breeding appears to be from three or four months to a year.

There are a number of records, however, of virgin females breeding at two years, and records of males breeding between three and four years. Extensive data show that for pairs which have produced a large number of litters the average litter-size remains reasonably stable up to the sixth or eighth litter, and declines rapidly thereafter.

The decline results from both reduction of the number born and from increase in the pre-weaning mortality. The majority of all matings have been accomplished by placing one pair in a cage and leaving the male in the cage for the duration of the mating period.

This system is used because it is difficult to detect early pregnancy and, in the experience of the writer, when two or three dams are used to a single sire the total progeny resulting is less numerous than would have been expected from a comparable number of females in single-pair matings.

The gestation period of a virgin or non-lactating female is 22 or 23 days; as in the laboratory mouse, there is a post-partum oestrus and the following gestation period is usually prolonged if the mother is nursing more than one or two young.

All mated pairs were checked at least weekly for the presence of pregnant females. Any cage found to contain a pregnant female was temporarily marked with a coloured tag, and these cages were then checked daily for the presence of young. When a newborn litter was found the appropriate entries were made in the records and the cage was marked with a tag of another colour.

Peromyscus maniculatus usually give birth to young within two or three hours of noon. Weaning can be safely accomplished as early as at 22 days of age. In the particular management system in use in the Ohio State colony, there was an inspection of the cages on the 22nd day after birth; if the young mice were robust, or if the delivery of the next litter appeared imminent, the young were weaned.

Otherwise they were left and observed daily; litters were always weaned by the 28th day. Although most of the strains maintained had been in the laboratory for a number of generations, the species is somewhat sensitive to day-length. One of the three animal rooms was equipped with a time-switch for light-control and this was used to provide a 15-hour day.

DISEASES

We have observed few cases of diseased animals, and no cases of ectoparasites; no examinations were made for endoparasites. The lack of ectoparasites is probably due to the use of cedar shavings for bedding. Individual cases of disease were not investigated and, unless a diseased individual was valuable, it was chloroformed and its cage mates were kept under careful observation.

The most common ailment we observed was a partial, sometimes complete, loss of hair for a brief period. This was often accompanied by a scabby appearance of the skin. Affected animals were isolated for two or three weeks and all but the most advanced cases would recover without further attention.

A condition that is relatively common in the *P. polonotus* stock is a slow but progressive loss of the tail, usually accompanied by a swollen, scabby knob at the end of the tail. Isolation of these mice does not retard the progress of the condition.

When the external tail has disappeared, the mouse continues to live normally and breeding is not prevented, although there appears to be a depressing effect. The same condition has been noted in *P. maniculatus*.

THE MOUSE

Surveys by the International Committee on Laboratory Animals have shown that in most countries where the information is available the mouse is the laboratory animal in greatest demand; from 40 to 80 per cent of all animals used in laboratories are mice. Cheapness, ease of breeding and maintenance, and a vast accumulation of experience with this species probably account for its use to such an extent.

The laboratory requiring mice either may obtain them from outside sources, such as commercial breeders, dealers, or other laboratories, or may breed them. The choice of one or other of these methods must be left to the user, as it is not yet clear which suffers from the greatest disadvantages.

In some cases the laboratory will be obliged to breed its own animals because its requirements are very specialized. Mice used in genetical and in certain types of cancer research must obviously be raised under the very close supervision of those actually using them.

Similarly, work involving the use of litter mates or young suckling mice necessitates the production of these animals where they will be used. In some countries all the animal requirements may have to be met within the laboratory's or institute's own facilities; this may not be a serious disadvantage, but it is doubtful whether any laboratory can produce mice as cheaply as commercial breeders can.

The user must balance various factors before deciding to obtain his supply by one means or the other. The bought mouse is usually cheaper, and arrives when required. There are no worries associated with breeding programmes; the very large commercial breeder can frequently cope with sudden big orders, and capital costs are lower.

The home-bred animal, however, has a known history of breeding and disease, has been fed and reared under known conditions, and does not suffer the hazards of long journeys under poorly controlled conditions. Further, its age can be guaranteed, and any other details required can be collected. Finally, its quality can be directly controlled by the scientist himself.

The ways and means of producing laboratory mice are many; there are as many schools of thought on the subject as there are breeders but, since the first edition of the UFAW Handbook, it is probably true that there has developed a certain uniformity in some aspects of the matter, as for example, in the widespread use of cubed diets.

The methods described in this chapter are a synthesis of what appear to be the most commendable methods and techniques employed in a variety of colonies in England at the present time.

The wild mouse is sometimes kept in captivity for certain experimental studies. The *UFAW Handbook*, covers this variety in some detail, but it will also be mentioned briefly in the present chapter.

ACCOMMODATION AND MAINTENANCE

Foundation stocks of mice can be fairly readily obtained from a variety of sources, and the Catalogue of Uniform Strains (uniform, not necessarily inbred) published by the Laboratory Animals Centre should enable one to find stock which conforms to one's personal requirements. Breeding nuclei of inbred strains are obtainable from a number of laboratories, details of which are given in a later section.

It is a wise plan to examine foundation stock very carefully for the presence of infection or other undesirable characters before embarking upon a full-scale breeding programme. A number of pairs of mice should be mated and the offspring of such matings should be used as the actual foundation breeding stock only if the parents, when killed after the young have been weaned, show no signs of infection at autopsy.

This post-mortem examination should be as detailed as facilities will allow. Particular attention should be paid to the various internal organs for signs of lesions or enlargement, and bacteriological cultures can be prepared from liver, spleen, gut, lungs and heart blood.

If any signs of infection are revealed from either parent, their offspring should be discarded. The more thorough one is at this stage, the greater the chance of founding a satisfactory colony.

Should an already established colony be infected, similar measures might be employed for the establishment of a new, clean, unit in some other room or building, situated as far from the old unit as practicable. For the production of SPF and germ-free animals the reader is referred to the other chapters of this book.

The Animal House

Little need be said about animal-house design since the factors involved are those common to the design of most small-animal houses, and have been discussed in an earlier chapter. One point, however, may be emphasized.

Mice are small animals, and very many can be kept in a small area. This condition is liable to precipitate outbreaks of epidemics, so that it is advisable to break the colony up into a number of small units housed in separate small rooms.

These rooms should be more or less self-contained, and organized so that should an outbreak occur individual rooms could be effectively isolated from the rest. Different strains should be kept quite separate.

Water points for filling bottles, and a hand basin, should be in each room, and heating provided. Good lighting should be installed but windows are not a necessity. Mice will breed well under a controlled artificial light cycle of 12 hours on, 12 hours off.

Mice need to be kept at a reasonable temperature to thrive, though the exact temperature is probably of less importance than its constancy. Large fluctuations should be avoided. A temperature of 18 to 24°C (about 65 to 75°F) is adequate, and draughts of cold air should be excluded.

In small animal houses the expense may not justify the installation of mechanical ventilation systems, but in any large unit these are worth while as they afford greater control over the animals' environment, especially where rooms are large or poorly situated for natural ventilation.

Full air-conditioning is to be recommended if it is at all possible; in temperate climates it may still be regarded as a luxury, but in tropical climates it is essential. No mouse colony will fulfil its potential in the climate of southern Indian, for example.

Airconditioning will give control over temperature, dust filtration, humidity, and the rate at which air is changed. With either full conditioning or just mechanical ventilation at least six to twelve changes per hour will be required.

The recirculation of air is not to be recommended, although under tropical conditions it may be necessary on the grounds of economy. If windows are used they should be flyproof and situated out of the noonday sun, and they should not expose animals to gusts of cold air when open.

Humidity is very rarely given much consideration in this country, and under normal British conditions it is unlikely to be of great concern to the average mouse-breeder; but excessive heat and high relative humidity, such as might very occasionally occur during the summer in poorly ventilated rooms, have been known to result in the loss of numbers of mice on rare occasions.

Adequate ventilation, however, should solve this kind of problem. An optimum relative humidity is about 50 to 60 per cent. Reference should be made to Lane-Petter (1963) who discusses the housing of mice and rats in some detail.

Cages and Racks

Mouse-cage designs are numerous, ever changing, and more or less unstandardized. Every colony is likely to use its own design and to consider it better than any other. Cagesize will depend upon several factors—the area given per mouse, the number of mice the cage is to hold, the purpose for which the cage is to be used, and ease of handling. Four classes of cages may be noted—breeding, stock, experimental and travelling—and each will he discussed separately.

For the small colony, by dint of judicious selection (assuming transport questions do not arise) only one standard cage might be used as an all-purpose cage, but for the larger colony it is frequently convenient to have three or four types.

It is difficult to define the amount of space needed by a mouse, but average values for this country are approximately 40 in^2 (250 cm^2) per mouse on experiment. These values would appear to be adequate. For a discussion of cage-sizes, see Tuffery (1955).

Cages can be made from a variety of materials-aluminium and aluminium alloys, galvanized iron, wood, tinplate, plastics, glass and stainless steel. Stainless steel is the most durable but it is extremely expensive and heavy to handle, and cages prove an expensive loss if their design is found to be at fault, or if a change-over in husbandry methods is contemplated.

Plastic cages are becoming more popular, being already in wide use in the U.S,A., and are now being developed in European countries. They are light, cheap, easy to clean and aesthetically attractive, and may be transparent or opaque.

The best are made of polypropylene, polycarbonate, or some other material capable of withstanding autoclaving. Some materials may be rather brittle, e.g. polystyrene, and thermoplastic materials are not so readily sterilized.

Figure elsewhere in this book shows a polypropylene breeding-box designed to hold one breeding pair and used at the Laboratory Animals Centre, Carshalton. It measures 12½ × 6 × 42½ in (31.5 × 15 × 11 cm). Figure elsewhere in this chapter shows a clear acrylic box measuring $11^{11}/_{16} \times 7^{7}/_{16} \times 5$ in (29 × 19 × 12.5 cm).

Plastic cages are likely to replace many other types in due course. A German design has been developed and tested by Spiegel and Gonnert (1961). Extra cheap, lightweight, non-autoclavable plastic cages are in use in the U.S.A. and are at present being tested in the U.K.

These are designed to be used once and then destroyed. Such disposable cages would appear to have advantages in certain laboratories, e.g. in a virus titration test for the determination of an ED 50, where deaths are counted over a short period.

At the end of this period all survivors are killed and discarded with their cage. Alternatively, the use of such cages is of value where highly infectious agents are being investigated, since they dispense with much cage and animal handling. Glass jars are often used for short-term experiments, but are otherwise not recommended because they have a high casualty rate and may be uncomfortable for the animals.

Wood is not recommended-it is liable to warp, the edges become rough, and the corners weaken unless waterproof glues have been used. Wooden cages should never be used for housing infected or other experimental animals. Galvanized iron has been the material of choice for many years and still has much to commend it.

The chief disadvantages of this material are its weight and the fact that regalvanizing is necessary after a time; mouse urine is a surprisingly effective corrosive agent if given long enough to act. Electroplated zinc is a good alternative to galvanizing. Aluminium and aluminium alloys have become popular as materials for all kinds of animal cages.

They are light and easily cleaned, do not corrode readily, and are pleasing in appearance. Alloys are preferable, and a reasonably heavy gauge should be chosen since all cages are subject to considerable wear and tear. Like plastic cages, aluminium alloy cages should have no exposed edges or ridges which give mice a purchase on which to chew. On some alloys an anodized finish gives a good, durable surface.

Tinplate is not recommended; in an emergency biscuit tins 9 in (22.5 cm) square and 5 in (12.5 cm) deep may be used, only as a temporary expedient. The lids are punched to provide openings for food hopper, bottle and ventilation. They are cheap, but their useful life is not long, and they should be replaced at the first opportunity.

Two basic designs of cage are in use nowadays-the separate or individual box type, and the battery or drawer type, and each type has its particular advantages. Besides durability, cages should be easy to handle and inspect, be simple to clean, stack well when not in use, and be provided with food hoppers and water bottles; only in exceptional cases should dishes be used.

As far as cleaning is concerned, the fewer joints, corners and ledges and the less wire mesh the better. All corners should be rounded and, in general, the more simple and streamlined the box, the easier it is to clean thoroughly.

Empty cages occupy valuable space when not in use, and this should be borne in mind when they are being ordered. The Glaxo cage shown in Figures elsewhere in this book is made of aluminium and measures 12 × 10 × 4 in (30 × 25 × 10 cm).

These cages do not stack easily but otherwise are satisfactory in use. The cages used at Harwell, also, are made of aluminium; they measure 10 × 8 × 6 in (25 × 20 × 15 cm) and are adapted to be conveniently nested for storage when empty.

Stock-Mouse Cages

Stock-mouse cages can be fairly large; the one shown in Fig. 4 measures 22 × 18 × 6 in (55 × 45 × 15 cm). It is made of anodized aluminium and has been divided into four sections by removable partitions made of thin-gauge stainless steel, each section housing from 15 to 20 mice according to their size.

The divisions were found necessary in order to prevent mice from heaping themselves in the corners and suffocating the lowest layers. A somewhat larger stock cage measuring 30 × 18 > 6 in (75 × 45 × 15 cm) may hold up to 100 mice if partitioned to prevent the mice from heaping themselves in the manner mentioned above.

On the ground of hygiene more crowded cages are not recommended, and stock from different sources should never be mixed in these cages. In the smaller colonies it is better to use standard breeding-cages for stock animals.

Cages of a battery type conserve space. The one used at Allington Farm and shown in Figure elsewhere in this book comprises 80 lidless drawers of the form shown in Fig. 6. Each drawer measures 18 × 5 × 5 in (45 × 12.5 × 12.5 cm) and is made of aluminium.

The racks are made of aluminium and the drawers of aluminium alloy. Although the initial outlay for such units might appear heavy, they are in the writer's opinion a good investment. They are most economical of floor space, and where a high production on small premises is needed, they are probably the most satisfactory system to use.

The racks are easy to move, being mounted on castors. Each cage can be readily inspected, and hoppers and water bottles are accessible with very little disturbance to the adjacent cages.

The size of the standard model is about 4 ft 9 in high, 4 ft 9 in long, 1 ft 6 in deep (1.44 × 1.44 × 0.45 m). The 80 drawers are housed on eight shelves in each rack, so that well over 100 weaned mice per week should be got from one unit. Double-sized units, equivalent to two units fixed back to back, can also be obtained.

There are no special requirements for mouse-cage racking and shelves, save that they must be adjustable to suit the cages in use.

Breeding-Cages

These can be fairly small, but the exact size will depend upon the breeding system in use; 18-in (45-cm) battery drawers will hold one or two does with their litters, and a buck.

The Harwell cage holds a pair or trio with litters, or six adult mice, and the Glaxo cage holds either one buck and four does with litters (a higher than average number), or two does with litters, separated by a partition. Average values indicate that 30 to 40 in^2 (200 to 265 cm^2) should be allocated to each doe with her litter, with nearly as much again for the buck if he is kept in during nursing.

Philip (1957) used a two-compartment wire-mesh cage for breeding wild mice, in which an upturned flowerpot provided a nesting box and retreat. Hay and sawdust were used as bedding and litter.

Transparent versus Opaque Walls

Mackintosh and Chance (1965a) have compared birth-rates in cages of three types, in two of which transparent plastic material was used while galvanized iron was used in the third type. The birth-rates in the plastic cages were found to be greater by 27 and 29 per cent, respectively, than the birth-rate in the galvanized-iron cage.

Experimentatl Cages

The design of cages for special purposes, such as metabolism and radio-isotope work, will often have to be left to the experimenter himself, and those concerned with this kind of cage are recommended to consult the following: Henry (1952-3)—a useful, if short, review of metabolism cages; Lazarow (1954); Ness *et al.* (1954)—three home-made metabolism cages; Charles (1954)—a good discussion of radio-isotope animal techniques; and Catchpole (1955)—a Perspex rat cage.

For most other types of experimental work, e.g. assays, toxicity and virulence tests, small cages housing up to 10 mice are generally used and special onemouse cages can be designed if these are needed frequently. Experimental cages must be capable of withstanding repeated sterilizing procedures.

Wood should never be used, and glass jars are not recommended. Battery cages of drawer type only 12 in (30 cm) deep may may be used. Experimental cages should be autoclaved at the end of the experiment before bedding and so on is removed.

Travelling-Cages

If mice are transported in one's own vehicles, as often happens at breeding-stations attached to a large institute or firm, it is frequently possible to use a cage or box of comparatively light construction since travelling conditions (i.e. heat and draughts) are under one's own control.

For rail and air travel stronger, non-returnable, boxes should be used. Boxes holding larger numbers of mice should be subdivided, each compartment holding about 20 to 25 animals.

Three or four square inches (20 or 26 cm^2) per mouse should be allowed, depending upon their size. Bedding should be provided-sawdust or peat to absorb urine and keep them reasonably clean, and wood wool to act as insulation against temperature-fluctuations and the shocks to which the boxes are bound to be subjected.

The boxes should be clearly labelled '*livestock*', and the address of both consignor and consignee .should be prominent. Cardboard travelling-boxes with ventilation holes covered with zinc mesh are cheap and serviceable for short inland journeys.

Boxes 13 × 9 × 4 in (33 × 23 × 10 cm) will hold small numbers of mice, boxes 14 × 12 × 6 in (35 × 30 × 15 cm) will hold fifty 12-to15-g mice, and boxes 18 × 18 × 9 in (45 × 45 × 22.5 cm) will hold 100 adults.

Animals travelling longer journeys (e.g. abroad) should be packed in stout boxes of wood or metal. The L.A.C. regularly dispatches mice (up to 6 or 7 per box) in 7-lb biscuit tins measuring 9 × 9 × 5 in (22.5 × 22.5 × 12.5 cm) and punched with air-holes. All travelling-boxes should be non-returnable.

The Cambridge Mouse Cage

Since the present chapter was written an important advance in the design of cages has been made by Dr M. E. Wallace of the Department of Genetics, Cambridge, who took as her starting-point the mouse's wishes and convenience, as deduced from behavioural studies.

It has turned out that the cage designed in this way also suits the convenience of the technicians and scientists who are responsible for the mice. For an illustrated description the reader is referred to Wallace (1965).

The cage comprises a shallow plastic box covered by a wire grid. The grid slopes downwards for about five-eighths of its length and then sharply upwards, forming a troughshaped depression in which pelleted food can be placed.

The remaining three-sixteenths of the grid is horizontal. The water-bottle lies on the long slope, parallel to the gradient, and next to one side of the box; it is furnished with a drinking-tube which projects through the shorter, sharply-sloping side of the trough-shaped depression.

Thus a narrow space, covered by open wires, is left under the horizontal part of the grid, and it is in this space that the mice eat and drink.

In cages of this type the mice have hitherto tended to build their nests under the supply of food and water, with the result that the food and nests become fouled; moreover the mice pile up sawdust until it reaches the drinking-tube, causing water to leak out and wet the floor of the cage.

The motive of this behaviour seems to be a desire to form the nest inside a solid shell and to exclude air except for an entrance hole; for this reason the mice have chosen to build their nests under the lowest part of the grid.

In order, therefore, to attract the nest away from the supply of water and food, a sheet of metal is laid on that part of the long slope of the grid which is not occupied by the water-bottle. The mice then build up their nest under this plate.

To eat and drink they wriggle under the water-bottle into the space under the horizontal part of the grid, and in this way sawdust is kept away from the drinking-tube and food, while food, urine and faeces are kept away from the nest. Cleaning of the cage is facilitated, and with automatic watering the time required for servicing is reduced to half a minute per cage per week.

Dr Wallace has also studied the optimum distance between the bars of the grid in order, on one hand, to enable mice to gnaw the pellets easily and, on the other, to prevent young mice from wriggling out or getting their heads jammed between wires.

Bedding

Bedding materials serve to absorb moisture, and provide insulation and nest-building materials. Sawdust and wood shavings are most frequently used, but peatmoss is another suitable material. Wood shavings are the least dusty of these alternatives and will be used by the mice for constructing nests.

Woodwool and clean shredded paper also make excellent nesting materials. Peat is usually reasonably clean, but sawdust can constitute a source of infection if it has been stored at some time in open yards where it may have been contaminated by wild rats, mice, and cats.

If one is not satisfied with one's supply of sawdust it is as well to inquire into storage conditions at its source as well as in one's own animal house. It should be dry, not green or dirty, and not too fine or dusty. Hardwood sawdusts may be harmful because of the presence of phenolic substances; soft or whitewood sawdust is best.

Woodwool (grade No. 1 in the U.K.) is clean, but the outside of the bales may be contaminated by rats and mice during storage; treatment with a blow-lamp should help to reduce the risks of introducing infection in this medium.

Where sterile bedding materials are required (for SPF stocks, for example) some care is needed to achieve this sterility. Gamma radiation is probably the best sterilizing medium but is unlikely to be available except under special circumstances.

The best alternative is probably the autoclave, but great care is needed to ensure the penetration of heat and steam to the centre of any bulk or pack of bedding-material. Most of these materials are poor conductors of heat, and it is important not to pack them too tightly nor at too great a depth.

In any autoclave run, test apparatus, such as Browne's tubes or sterilizing tape, should be packed in the part of the load likely to be most difficult for the steam to penetrate. Highspeed high-vacuum autoclaves are the best for such loads. Ethylene oxide is a good sterilizing medium for animal bedding, but at present its use for this purpose is still largely in the development stage.

FIGHTING

The following note on fighting and its prevention has been contributed by Mackintosh and Chance (1965b). Mice which are strangers to one another, either because they come from different groups or because they have been kept solitary for a few days, fight.

The intensity in general increases with the duration of the isolation, and when the fighting is intense they will inflict severe bites which can become infected. Accordingly precautions should be taken, when introducing strange mice to one another, to reduce the severity of fighting.

Fighting is almost always a male activity and is most intense between males, but may also be directed at females. It may be intense at the start, but usually declines after the first hour or so.

Experiments on the olfactory stimulus to fighting, provided by the smell of other mice, show that an important, or perhaps the most important, factor causing one mouse to attack another is unfamiliarity with the other's smell. It might be possible to control the amount of fighting by masking the smell.

The outbreak of fighting, however, involves recognition by one mouse of visual stimuli from the other mouse and, if these are obstructed by adding some cover material when strange mice are mixed in groups, fighting is considerably reduced. For example, a handful of woodwool placed in the cage has been shown to reduce fighting between male strangers by between 50 and 75 per cent.

WATERING

Open water pots should never be used, any more than open pots for food. Bottles, not glass bulbs, are to be preferred. The exact design of the bottle will depend upon the design of the cage, particularly in the case of battery cages where the bottle has to fit into a specially designed niche.

They can be of almost any convenient size. Plain, round, wide-necked bottles are easiest to clean, either by hand or mechanically. The stoppers can be either screw-on caps with metal tubes, rubber bungs with metal tubes, or plastic with metal tubes.

Screw-top bottles are slightly more laborious to open, but less likely on this account to open inadvertently and swamp the cage. Screw-caps of good quality, or well-fitting bungs, should be chosen. The cheapest bottle, which is quite satisfactory if treated well, is one with a rubber bung and metal tube, and the most costly is one with a plated brass cap, metal tube, and rubber washer.

Both types find favour. Autoclavable plastic bottles are becoming popular; their breakage rate is low but their initial cost is high. Glass tubes are not so safe to use as metal tubes. In any type of bottle the spout or tube should be long enough for the animals to reach, but not so long as to touch the bedding and drain the water away.

One can use either a large bottle needing attention only once a week, or a small bottle needing refilling on alternate days. Bottles should preferably be changed rather than topped up, but if bottles must be refilled they should always be returned to the same cage.

Many portable bottle-filling taps are available nowadays, although they are readily made from standard water guns, such as that described by Tuffery (1957).

NUTRITION

'The prerequisite of all experimental work with laboratory animals is a stock diet upon which the stock colony will thrive and reproduce itself efficiently'. There should be no need to emphasize this fact any further, but difficulties arise when one compounds, and then attempts to assess, diets.

The nutritional needs of the mouse have not been so widely studied as those of the rat, although there is a considerable literature on the subject, and those especially concerned are referred to the Various relevant journals and reviews. Morris (1944) has reviewed the knowledge of mouse nutrition up to 1944, and Russell (1948) that up to 1948. Worden (1953) gives a number of additional references.

A recent review by Porter (1963) covers the ground very thoroughly. Observations by Porter show that widely accepted diets, which are commercially available in the U.K. are not always as reliable as one might suppose.

He found that batches of Diet 41B made to the same formula by three different manufacturers gave widely different results in breeding trials. Furthermore, these trials needed to be carefully designed to yield unequivocal and useful results.

In spite of these observations, the use of cubed diets constitutes a great improvement over older methods of feeding, but such diets must not be used uncritically or relied upon absolutely.

The feeding of mashes is becoming less and less frequent and the method (unless exceptional circumstances demand it) is not to be advocated. Mashes are messy, unhygienic, laborious to prepare, liable to go sour easily, and not readily stored.

Cubed diets are widely used in all types of colonies nowadays, and mashes are rapidly losing favour. In this country three formulae are more widely used than others: Diet 41; Diet 86; and 'Thomson Rat Cubes'. All three are available commercially.

In America several firms market cubed mouse diets, e.g. Rockland Rat and Mouse Diets; a French cubed diet is Sorolabo. Laboratories in countries with no large established animal-foodstuffs industry may need to investigate and develop their own diets which are prepared from locally available cereals. These diets can be cubed, if desired, on small cubing machines which are available in a number of European countries and in the U.S.A.

The Diet 41 designed by Bruce was evolved as a complete diet needing no supplements, and Diet 86 is described as 'complete'. Thomson's diet, when fed to rats, was intended to be accompanied by a small portion of kale anti separated milk.

A modified Thomson diet described by Gruneberg (1943) has been used by Philip (1957) for laboratory-bred wild mice. Supplements are frequently given with these diets, but in some cases, at least, custom rather than necessity is the reason for this.

It seems very likely that different strains of mice do have different nutritional requirements, but at the same time a plea for the less haphazard approach to the question of mouse foodstuffs might reasonably be entered here. A good instance of inadequacy in batches of diet of a standard formula is given by Blackmore and Williams (1956).

In this case strain of mouse, time of year, batch of diet, and formula of diet were of undoubted relevance in explaining the reduction in productivity described. Dietary deficiencies may manifest themselves after (or during) gestation, parturition, or lactation, or even in the third generation fed on the deficient diet.

Porter et al. (1963b) found that dietary insufficiencies revealed themselves in poor or reduced breeding performance and poor weight-gains. Animals are quite liable to resent changes in their accustomed way of living, and a sudden change of diet may result in loss of weight and of general vigour, or even in more drastic effects.

Mice accustomed to meals of kitchen scraps have been known to refuse the more plebeian cubes when sent from one breeder to a purchaser's colony. Where changes in diet are contemplated it is wise to make the process a gradual one.

Texture may also influence the acceptability of a foodstuff, and storage resulting in the souring of one or other component may affect a diet's palatability. All animal diets comprise essentially a mixture of substances which are at present uncontrollable as to food-value and quality-wheat, barley, fishmeal etc.—and this may account for the apparent variation between batches.

Despite these difficulties, however, diets conforming to a published formula represent a valuable step forward in the production of animals of uniform good quality, and the three named diets popularly in use in Britain are to be recommended.

Cubed diets for mice are best offered in hoppers. These are commonly of wire mesh, though punched aluminium alloy has been tried with success. The holes through which the mice chew the food should be only as big as is necessary to enable an adult mouse's jaw to work.

In punched-metal hoppers either large circular holes or smaller but vertically elongated holes should be used. Hoppers reduce the fouling of food, cut wastage due to scattering, and are very easy to refill. They can be made separate for ease of cleaning, and suspended from the lid of the cage. A hopper of this type is used in the Glaxo cage, but in the

Harwell and L.A.C. boxes the hoppers are built into the lids. Hoppers should, of course, be within reach of the mice, but not so near the cage floor as to touch the bedding.

BREEDING

The physiological processes involved in reproduction have been studied in some detail in mice, though to not quite the same extent as in other animals. This is not the place to embark upon a discussion of the biology of mouse breeding and the interested reader is recommended to consult other sources.

The reproduction chapters in *The Biology of the Laboratory Mouse* by Snell (1956) give a useful review of the main aspects of the problem and include over 200 references up to 1941; see also Bruce (1963), and Bateman (1957) who deals with the physiology of lactation.

The complete oestrous cycle in adult females occurs about every five days in unmated animals. There is usually a post-partum oestrus within 20 to 24 hours of parturition in the mouse, followed by a long period of dioestrus during lactation lasting 20 to 25 days.

Weaning induces oestrus within two to four days. Strains vary in their rate of development, but generally mice are sexually mature at about 6 to 8 weeks of age. Weights at birth, weaning and sexual maturity vary considerably from strain to strain, so that no common standard can be given.

For the L.A.C. Grey strain bred at Carshalton these figures are only approximate: birth 1.0 to 1.5 g; weaning 10 to 12 g; maturity: males 28 g, females 25 g. Weights at birth and weaning are both influenced by litter-size, and the latter by age at weaning also. The *Catalogue of Uniform Strains* lists data of this type for a variety of strains, but there is no measure of the accuracy of the figures given by the various authorities.

Litter-size varies considerably, inbred strains giving in general smaller litters than noninbred mice. This number also varies with parity, age of parent, and other factors, but for the average non-inbred colony it should be at least seven or eight, and higher averages can be aimed at in some colonies.

Gestation normally lasts about 20 to 21 days, but where a fertile mating took place at the post-partum oestrus, this period may be lengthened by seven or more days, the lengthening depending partly upon the size of the suckling litter. This again seems to vary with the strain.

Advantage can be taken of the post-partum oestrus for practical breeding programmes, and two methods will be described in detail, one using, the other not using, post-partum mating. (See the section on inbred mice for variations of these mating procedures adapted to the requirements of inbreeding.)

Monogamous Pairs

One male is mated with each female under this system, and they are left together for the rest of their useful breeding life. Mating may be effected at any age between 8 and 12 weeks, 10 being a good average. It is doubtful if any advantage accrues from earlier or later mating.

Three to four weeks after mating one can expect the birth of the first litter, and thereafter, since in the majority of cases mating occurs at the post-partum oestrus, a litter can be expected roughly every third or fourth week. With this system weaning must be carried out just before the next litter is expected, even if at times the young appear to be a little smaller than one would like.

One can also use a trio system, permanently mating one male with two females. Increasing the number of females beyond two, results in a rapid fall in the numbers weaned per female, although often the number of mice weaned per box per unit time may continue to rise.

This last consideration may lead to the choice of a one-male, four-females system where space is at a premium and one is concerned to get the maximum production out of the smallest area.

One penalty for such a scheme can be greater variation in age/weight ranges at weaning (i.e. on removal from the parents' box), and this could be of importance in some experimental work. In spite of this, this mating system is widely used.

Harems

Under the harem system up to four does are regularly mated to one buck, but as soon as pregnancy is definitely established the doe is removed to a separate cage and allowed to litter down and nurse the young on her own. Under this system each doe will have a litter only once in seven or eight weeks, being returned to the buck at weaning.

Weaning, however, need not take place at any limited time, and consequently the average weight at weaning under this system can be greater than under the pairs system.

The choice of mating system will depend upon the requirements of the laboratory using the animals, the space and other facilities available, the staff available, and other less obvious factors, the most important of which concerns the characteristics of the strain of mice being bred.

With some strains, for example, the presence of the male at parturition and nursing leads to excessive losses by neglect or cannibalism. On balance, the writer favours a pair or trio system, on the grounds that such systems take less labour and cages, yield litters more frequently and, because they involve less handling and movement of the animals, are safer on hygienic grounds.

Contrary to expectations, it has been shown that the harderworking monogamous pairs had a better health record than harem-bred animals in one institute. Lane-Petter (1955) and Bruce (1947, 1954) investigated the merits of different mating schemes; other laboratories might well institute similar investigations before regarding the current established method as undoubtedly the best for their purposes.

There are a number of published descriptions of methods of breeding mice which might be usefully consulted. A very good American paper by Poiley (1952) considers a number of important aspects of the problem. A chapter in a book by Gruneberg (1943) will be of interest to geneticists. Others include Eaton and Cabell (1949); Keeler (1931); the mouse chapter by Strong (1950) in *The Care and Breeding of Laboratory Animals;* Dekking (1952), Porter (1951); Isaacson (1952); Hoyland *et al.* (1963) and Wallace (1963).

Since the early work on the use of unweaned mice for the titration of foot-and-mouth virus unweaned mice have become an extremely useful tool for virologists. The production of small numbers of such animals offers no great difficulties, but the organization of a large supply of unweaned animals requires careful planning.

Originally, pregnant females were bought as required by Skinner (1957) but it soon became necessary to set up a breeding colony proper. The organization and history of such a unit has been recorded by Davis (1960), Davis and Hoyland (1961), and Hoyland *et al.* (1963).

These articles should be referred to for details, and Skinner (1953, 1957) gives an account of the use of sucking mice for work with the virus of foot-and-mouth disease. Dr Skinner has also kindly contributed the following notes on the fostering of unweaned mice, as a means of reducing the consumption of breeding females in the production of foot-and-mouth vaccine.

'Unweaned mice infected with modified strains of foot-and-mouth disease virus have been used by the Research Institute, Pirbright, as a source of tissue for the large-scale preparation of live vaccines for cattle (Hollom *et al.,* 1962).

For these particular mice, a system of using foster mothers was adopted in order to reduce to a minimum the drain of females from the breeding stock. This procedure was quite feasible with the mice of the random-bred Pirbright P strain colony.

Since 1950 this colony has been made the main source of unweaned mice used routinely at the age of 4 to 7 days for the titration of foot-and-mouth-disease virus. Before their use for this particular purpose large groups of young are randomized and coloured and are then redistributed among the mothers.

The incidence of cannibalism as a result of these and subsequent manipulations is very low. 'During the large-scale production of live vaccine the procedure used was the following. Mice, 8 days old, from the P strain colony were inoculated intraperitoneally with a virus dose chosen to produce a fatal infection to the mice in 20 to 23 hours.

At the 23rd hour the dead mice were collected from the tins and they were replaced by the next batch of inoculated mice. This was repeated on four successive days. On the fifth day the mice substituted were uninoculated and only five days old.

These mice were inoculated three days later to commence a second cycle of infections in the following week. The same mothers were used again finally for a similar cycle in the third week. The fostered mice showed no reduction in virus sensitivity.

'The number of female mice withdrawn from the breeding stock was in this way reduced by over 90 per cent. Each mother was responsible for carrying through 12 litters and could have been carried through at least six more if inoculations had continued daily throughout the three-week period.

The mothers of the infected unweaned mice were used in the breeding unit as additional mothers for newly born mice. This improved the condition and weight of these mice before their issue for virus production and resulted in a higher yield of infective tissue. These mothers later became available for re-mating for further production of litters.'

Young mice, newly weaned, frequently need holding for a week or two before issue at the required age or weight. The general practice is to group them together in larger stock boxes in mixed groups of 10 or so but it is a better plan, where space and cages permit, to keep each litter entire in smaller cages until ready for issue.

This is particularly valuable for tracing disease back to individual breeding animals—should one or two stock mice become sick or die, the parents can be traced and culled if necessary.

Supply of Inbred Mice

The Standardized Nomenclature for inbred strains of mice, second listing (Snell *et al.*, 1960) gives an index list of inbred strains and the laboratories maintaining them. Some of these laboratories (e.g. the Laboratory Animals Centre in England and the Roscoe B. Jackson Laboratory in the U.S.A.) maintain stocks for the specific purpose of supplying small breeding nuclei for expansion in other establishments.

The Roscoe B. Jackson Memorial Laboratory also has a Production Department for the more popular strains and can usually supply these mice in larger numbers for experimentation.

Most laboratories are willing to supply a small breeding nucleus of one or two litters from time to time but are usually unable to guarantee a regular supply.

Laboratories wishing to use inbred mice in this country usually find that they have to breed the animals themselves. Occasionally, if mice are required in small numbers or for a single experiment, it may be possible to arrange to take the surplus animals from another animal house; from time to time the Laboratory Animals Centre does have mice which can be issued for experimental purposes.

In the normal event of the user breeding his own animals he must decide whether to maintain his own inbred strain or subcultivate breedingstock from an inbred strain maintained elsewhere. For most purposes subcultivation of stock provides the easier solution to the problem of breeding inbred mice.

For some work, however, the exact relationships of the experimental animals must be known, and in such cases the user must have access to his own inbred strain. Inbred strains are valuable research tools and special techniques are necessary for their maintenance, which must not be left to untrained workers.

A full account of inbreeding, selection and maintenance of inbred strains is given by Falconer (1962) and Dinsley (1963).

Subcultivation of Inbred Mice

The L.A.C. subcultivation scheme is simple to use and has many advantages. The L.A.C. inbred strains are colonies of primary type designed mainly for supplying breeding nuclei for subsequent expansion in other laboratories.

By this method the establishment of many sub-lines of the same inbred strain is avoided, and the user laboratories are freed from the laborious work of maintaining their own inbred strains.

One colony of primary type is capable of supplying breeding-stock to many expansion units; the colony itself, however, remains small enough for the exercise of strict health-control and genetical-control. All laboratories obtaining mice from the same primary-type colonies are using animals of similar genetic constitution.

This means that valid comparisons can be made between experiments performed in these laboratories in so far as the results depend on uniform genetic constitution of the animals.

Under this subcultivation scheme mice are supplied from the colony of primary type and random-mated in the expansion units for three generations only, mice of the third generation being used entirely for experimental purposes and not for breeding.

New breeding-stock is drawn at intervals from the primary-type colony so that the subcultivated stocks retain their identity; three generations of random breeding are not likely to alter the characteristics of the strain.

With this scheme in operation large numbers of virtually inbred mice can be produced in the expansion units. Breeding-stock is supplied in pairs or trios (i.e. one male with two females). These mice should be random-mated within their generation irrespective of relationship; the male need not be removed at any time from the female or females.

The progeny of this original stock is the first generation and should be used for replenishing breeding-stock. This process can be continued until the third generation but no further. All mice of the third generation are used as experimental animals.

A simple method of distinguishing the generation number is by a traffic-light colour code. White labels are used for the stock originating from the Laboratory Animals Centre. The first, second and third generations are labelled respectively green, yellow and red.

Progeny from white boxes always go into green ones, those from green boxes into yellow, and those from yellow boxes into red. Mice in boxes with red labels are for experimental use only and must not be used for breeding.

The original white-label stock is replenished by the L.A.C. It is best in practice to maintain a steady turnover with replenishment at regular intervals, e.g. every six months, so that once the colony has been established there should always be breeding-stock representing each generation.

It is essential that the expansion units should follow the above method exactly if the strain is to keep its identity. Under this system, the sub-colonies in the expansion units will have the same designation as the primary inbred strain maintained at the L.A.C.

Selection

The breeding of mice must be as economical as possible, provided that attention is paid to quality as well as quantity. One factor over which some control might be exercised is that of selection for particular desirable features.

Obviously, health is of especial value and one does not continue to breed from animals with a family history of disease. Next to this, for the general-purpose colony, the main interest is in productivity. One wants the maximum number of mice consistent with a certain standard of health from the minimum number of adults in the minimum space of time.

For most practical purposes, then, an attempt will be made to select for, first, general health and vigour, second, high production and mothering ability, and finally for a variety of other factors such as tameness and ease of handling.

The two main causes of variation in animals are heredity and environment, and in the ordinary mouse colony one attempts to stabilize the latter to a satisfactory level and select for hereditary characters. Unfortunately, it is not always easy to distinguish one from the other, but with a little careful thought a sound scheme of selection can be introduced.

In practice a colony is run in two sections in the first are the breeding adults which produce the breeding mice for the second section, and these latter mice are the ones which actually produce the mice which are issued for experimental purposes.

Selection is made for productivity and mothering ability, i.e. one keeps, to supply the future breeders, those litters of mice which are above average in size and general health and vigour. In the first section (the breeder-breeder section) selection can be based upon a productivity index. Q, which is defined as the number of mice weaned per 100 pre-natal days i.e.

(number iveaned × 100)/ (number of davs since birth of previous liner).

In practice, this index is calculated cumulatively each time a litter is weaned. In the second, or production section. selection is largely a matter of culling infertile pairs or trios and obvious poor doers. The unselected litters are either issued for use or turned over to become breeders in the second section of the colony, as needs dictate.

When the selected litters have grown and reached maturity and are ready for mating, one selects the best individuals from each litter, having regard especially to health and disease-resistance, and mates these. The selection process is of course continuous, and for more detailed accounts of this topic, see Falconer (1952, 1962) and Dinsley (1963).

RECORDING

The keeping of records is a vital adjunct to the efficient running of any colony. Records may be as full as one cares to make them, but where time and labour are at a premium much useful information can be extrapolated from a very small number of data, provided these few are accurate and of the right sort.

It is necessary to collect information which will describe the performance of different sections of the colony and of the colony as a whole. A pairs system is the simplest to record, but it is not difficult to keep track of other systems, especially if attention is concentrated on the females.

Each female should always be mated with the same male, and in general it is easiest to judge the record of them together rather than each separately. For measuring the productivity of a breeding doe one will need to keep note of the dates of matings, date of birth of each litter, size of litter at birth, size at weaning, date of weaning, and parity.

To these basic data might be added still-birth numbers, losses before weaning, weight at birth and weaning, and notes on incidence of disease. Finally each animal, or the pair together, should be given a reference or index number and a note of their ancestry (parents' reference numbers).

This type of record can be kept in books, in loose-leaf notebooks, or on cards—the last is probably best as it allows most readily for the inevitable expansions and alterations. These cards can be kept attached to the cages themselves or in a separate cabinet; the latter alternative is better since the cards are less likely to be damaged, and the cage itself has nothing on it except a slip with an index number.

Besides the cards there should be a log-book in which the various items are summarized either daily, weekly or monthly, as desired. This record will show quickly such things as daily births and weanings, numbers of newly mated and discarded breeders, total numbers of deaths and culls, and balance of stock in hand.

The log-book record can be referred to when one wants to know, for example, how many breeding females are in production at any one time. It can also be used for extracting population data for various purposes. From the log-book output and productivity can be traced over regular intervals, e.g. total births!weans per week, and so on. This type of information, summarized at regular intervals, will furnish a means for following the progress of the colony.

For accuracy, the data should be tabulated, but simple graphs will illustrate the general trends most readily. Furthermore, such data provide the basis for cost analyses. The individual record cards will supply the information upon which selection is based.

Suppose one decides to use as breeding replacements the young from parents whose average litter-size (weaned) exceeds a certain arbitrary value (or, what amounts to the same thing, whose total number weaned exceeds a certain value), then it should be a very simple matter to design a card with a column in which is entered the cumulative total weaned, as well as one showing total weaned per litter.

If replacements are chosen on the basis of performance up to and including the first three litters, then selection of breeding-stock becomes simply a matter of using those animals whose cards show the required figures as equalled or exceeded.

Some breeders like to grade their does. The breeding performance is kept on an index card, and when the third litter has been weaned a coloured signal is attached, indicating the female's grade. The levels upon which grades are assigned are arbitrarily chosen in the light of one's experience of the strain of mouse in use.

After grading, only offspring from grade-A does (or perhaps grade-B does in an emergency) supply the breeding replacements, and grade-C offspring are used only to breed mice for issue.

These brief notes on recordkeeping may make the process seem complicated and time-consuming, but it need only be once organized and the returns, in knowledge of the colony as a whole and in improved breeding performance, are worth the trouble involved.

In summary, the system recommended involves keeping one record card per breeding doe, and a log-book showing total colony production at regular intervals. If a visual record is required as well, simple graphs can be constructed from data in the log-book.

HANDLING

Mice are relatively easy to handle and will usually bite only the hand that is afraid of them. For general purposes such as transferring them to clean cages and so on they can be picked up merely by closing the hand more or less completely around them, or by holding their tails.

For experimental procedures they are best held between the thumb and forefinger at the back of the neck—they must be held firmly so that they cannot twist round and bite, but not be strangled. After a little practice one operator can pick them up in the left hand in this way, and give them as intraperitoneal injection with the right hand without any assistance.

Wild varieties are extremely shy and will bite furiously when caught, but later generations bred in captivity become easier to handle. Intravenous inoculations are best done into the tail veins. The animal is placed in a tube held at a convenient angle, with its tail hanging out of one end.

It can be shut in firmly with a cark having a notch through which the tail protrudes. The other end of the tube can be blocked with a wedge of wire gauze. The veins can be dilated by dipping the tail in warm water before the injection is made.

Various strains will differ in behaviour and temperament, and these differences will soon be recognized by an intelligent technician. Some are excitable and some quiet; some squeak on the least provocation.

Some will breed best under a monogamous-pairs system, while others will wean larger litters if allowed to kindle and nurse in separate cages. The technician himself will influence the animals' patterns of behaviour, and one can observe interesting variations in such things as gain in weight and mortality which show links with the duty rota of technicians.

Lane-Petter (1953) observed that mice showed a slowing down, or even reversal, of the rate of gain of weight over the week-ends (Friday, Saturday and Sunday) as compared with the rest of the week, and Mundy (1954) reports an increased mortality among mice over the week-end period.

One would hesitate to offer an explanation of this kind of thing, but it is obvious that mice, and other animals, are not just reagents, but living beings which will respond to environmental influences. ft follows, therefore, that the environment, largely controlled by the attendant technician, should be a good one.

ANAESTHESIA AND EUTHANASIA

Ether is quite satisfactory for anaesthetizing mice for short periods. One convenient method is to place the animals in a 10-in (25-cm) desiccator into which ether vapour can be pumped; a rubber bulb can be used to blow air through an ounce or two (30 to 60 g) of ether contained in a small bottle, and then through an inlet tube passed through the lid of the desiccator. A less elaborate method is to place a wad of cotton wool soaked in ether under the gauze, in the lower half of the desiccator.

For killing mice, a similar procedure can be used by prolonging the exposure to ether. (Chloroform must not be used for anaesthetizing mice where there is a risk of its vapour reaching the main stock, as the males of some strains are extremely susceptible to small concentrations of this substance, dying suddenly from nephritis after such an exposure.)

Mice can also be killed by bleeding from the axilla under anaesthesia if one requires the blood for, say, serological purposes. Where only a very few mice are to be killed, they can be most easily dealt with by placing them in ajar of suitable size containing a pad of cotton wool soaked in ether or chloroform; they must not be wetted from the killing agent, contact of which with the skin is very painful.

For long periods of anaesthesia Nembutal, used in accordance with the makers' instructions, is recommended. Dislocating the neck is a simple and humane method of killing mice. The animal is held by its tail and placed on to a surface it can grip, when it will stretch itself out so that a pencil or similar object can be placed firmly across the neck. A sharp pull on the tail will then dislocate the neck and kill at once.

THE MULTIMAMMATE MOUSE

Adult males of this species weigh up to about 100 g. The head and body in the wild measure 115 to 135 mm, with a tail 112 to 120 mm long. Females weigh up to 80 g. The colour is something between those of the house mouse and of the black rat.

Anatomically, *Mastomys* resembles the rat, e.g. in the absence of a gall bladder, although careful comparison with both rat and mouse emphasized that it conforms satisfactorily with neither, and the presence of several unique features-notably a well-developed female ventral prostate, a cordate anterior process of the baculum (*os penis*), and the absence of preputial glands would justify its allocation to a separate genus or subgenus.

The yellow-brown colour of the incisor teeth is due to ferric ions (haemosiderin). The description 'multimammate' is due to the presence of 8 to 10 pairs of teats, with nipples more or less evenly spaced along the mammary line.

DISTRIBUTION

Praomvs natalensis is one of the commonest rodents in Africa, and its distribution extends from the Eastern Cape Province to Morocco. It is one of the few species to become partly domiciliated. It forms a link between human habitations and wild rodent populations nearby, because it makes use of the burrows of other rodents in preference to digging its own.

It is an important vector of plague, and precautions against escape are especially necessary. On the western seaboard of the United States, for example, it might prove a special menace.

LABORATORY ADAPTATION AND USES

The species was introduced to the laboratory in 1939 by Davis, from whom my stock was obtained in 1958. It has since become much more docile than it was, probably as a result of modification in caging and handling procedures. Its uses in the laboratory have been summarized by Davis (1963), to whose work the reader is referred.

Cancer

Its potential value in cancer research is considerable, for it is a museum of spontaneous tumours. The pattern of tumours varies with the laboratory and with inbred lines within a laboratory. Our stock has been uniquely susceptible to spontaneous carcinomas of the glandular stomach.

These gastric carcinomas differ from the human cancers in that they affect regions of fundic glands more often than pyloric glands, and are not preceded by intestinal metaplasia, but rather by a precancerous hyperplasia of the fovelae. In my stock, stomach cancers were commoner in females, but males preponderate in the stock of the National Cancer Institute.

In certain lines, animals over one year of age tend to develop papillomas of the skin, predominantly on muzzle and ears, and in females on the perineum. These appear as cutaneous horns which may attain 5 cm in length, or as fungating, crateriform or inverted papillomas. Larger lesions are often associated with underlying abscesses.

Perineal papillomatosis usually extends into the vagina, and here progresses into squamous-cell carcinoma more often than do papillomas of the skin. At present this tumour attacks approximately 95 per cent of the R strain by the time they reach two years, and is a common cause of death.

Other spontaneous tumours include lipomas, spindle-cell sarcomas, haemangiosarcomas, hepatomas, bile-duct papillomas, renal cortical carcinomas, granulosa cell tumours of the ovary, cervical polyps, uterine fibromyomas and haemangiomas, anterior pituitary adenomas, adrenal cortical adenomas and phaeochromocytomas, meningiomas, and a neuroma with granular-cell myoblastoma.

A lesion resembling *molluscum contagiosum* has been encountered in the wild. In my stock, primary hepatomas have been very rare except in association with severe liver damage caused by *S. mansoni*, which also induced reticulum-cell sarcomas in association with egg deposits. In other colonies hepatomas have been common. Granulosa-cell tumours of the ovary were common in three lines which have now died out, and have also been noted in other stocks.

GENERAL CHARACTERISTICS OF HEALTHY ANIMALS

The animals are moderately active by day, but far more lively at night. They exhibit quick nervous movements, either in attack or in evasion, usually a hop or jump followed by a run. They swim and climb well, and can climb a string.

Their nesting instinct is strong and they construct nests with any available bedding. Mated animals are cleanly and, in solid cages, push faecal pellets out of the air vents. With a wire-bottomed cage they may move off the paper bedding to urinate through the free wire, keeping this area for excretion.

These features are consistent with the finding in nature that the burrows are not contaminated with faecal pellets. Veenstra (1956) has noted that this cleanliness is not so obvious in cages containing a number of animals of one sex. Characteristically a colony has no detectable smell.

In Wood's light the hair of young as well as adult animals show a russet fluorescence, as does that of a number of other African rodents. Red fluorescence of the Harderian glands is absent.

Physiological Normals

The average rectal temperatures (obtained with a clinical thermometer inserted for 2 minutes) were 96.7°F (35.9°C) in 50 males, 98.4°F (36.9°C) in 26 non-lactating females, and 99.6°F (37.5°C) in 24 lactating females.

Like the mouse, *Mastor vs* concentrates iodine in the sub-mandibular gland and alkaline phosphatase is not histochemically demonstrable in the neutrophil leucocytes.

In the R strain at weaning, on the 21st day, males weigh 10 g and females 8.9 g. Maximum weight is attained by one year: in the R strain the average weights are 69 g in males and 49 g in females, which is less than those of outbred stocks.

The life-span in males may extend into the 4th year, but is generally under 3 years. Females do not attain 3 years. The statements by Oliff (1953) and Johnston and Oliff (1954) regarding age at reproductive maturity appear to have been gross overestimates according to Meester's as well as our own experience.

Cannibalism

Towards other rodents *Mastomys* is non-aggressive in the laboratory, as in the wild state. Veenstra (1958) suggests that its territorial instinct is very weak. In respect of its own species, on the other hand, it frequently exhibits cannibalism. This ranges from the normal eating of the placenta after delivery (males as well as females participate), to eating of litters, of dead cage mates, of living cage mates and even of themselves.

Cannibalism of the living mainly affects inbred lines. Litter-eating is common, especially with inbred lines and with first litters. In part this behaviour indicates defective progeny, for such a mother may accept healthy hybrid pups and foster them successfully, even though her own may be eaten: in part it indicates a defective mother, as the young are often strewn about, even out of the nest.

Litter-eating at one stage resulted in the loss of 96 per cent of all litters. Supplements of the diet with wheat germ, milk, and bread soaked in vitamin E failed. Litter-eating diminished steadily, however, as the inbred line progressed, which may reflect selection of the more successful breeders.

Occasionally cage mates are attacked and eaten. Sometimes this is confined to nibbling away the ears, the victim making no attempt at escape. It is commonest when males are caged together, next in females, and least in mated animals. It increases with age and at lower temperatures.

It is also more likely to occur when the victim is weakened by disease. Controlled experiments with the provision of blocks of wood to provide alternative objects for gnawing did not reduce cannibalism, nor did supplements of potato, wheat. germ, green vegetable matter (grass, lucerne and cabbage) make any difference. Total darkness was equally ineffective.

ACCOMMODATION

Cages are made of super-molybdenum stainless steel No. 1 (non-magnetic) and measure 6 × 6 × 12 in (15 × 15 × 30 cm). Walls are of 24-gauge sheet with bottoms of stainless-steel screening (18-gauge, 6.35 mm internal mesh), giving 3.3 spaces per in (1.3 per cm).

The lid is detachable and, apart from the rim, composed of standard mesh screening. It is perforated with a hole 3 in (7.5 cm) in diameter centred 3 in from one end, through which the food hopper is inserted.

This hopper is 3 in in diameter and 3— in (9 cm) deep, and is made of a flanged cuff of stainless-steel plate, 2 in (5 cm) deep to which is welded a cup of stainless steel screening of standard mesh, projecting 1½ in (4 cm).

Cage racks are 6 ft (183 cm) long and 27 in (68 cm) wide with three tiers having 3-in (7.5-cm) casters. A low partition divides each tier down the centre. Eleven cages can be accommodated on either side, resting on two longitudinal ½-in (12.5-mm) stainless-steel rods, 8 in (20 cm) apart.

A sheet of stainless steel 2 in (5 cm) below these rods supports a sheet measuring 6 < 2 ft (183 - 61 cm) and made of polyvinyl chloride on woven nylon fabric to collect the droppings. The spacing of the tiers is generous, but is designed to allow direct inspection of the cages through the mesh roof.

The lowest cages rest 12.5 in (32 cm) above the floor, the middle tier 19 in (48 cm) above this and the upper tier 17 in (43 cm) above that. The roof of the highest cage is thus 54[1] in (138 cm) above the floor.

This system is expensive in capital outlay and rather extravagant of space and in food lost through the floor. It is, however, more economical of technicians' time; we change cages every four weeks.

The room-temperature is maintained at 78°F (25.5°C) varying two or three degrees from top to bottom row. An extraction fan provides three changes of air per hour.

In our laboratory colony, thermostatic control does not prevent severe sudden drops in temperature during winter. The number of litters in which all pups are eaten is greater in the winter months, when weanlings have been obtained from approximately 50 per cent of litters.

Fighting

When strange animals are put together there may be some wild fighting initially, especially when males are so congregated.

Transport

To prevent escapes, which may be serious in plague regions, cages must be able to resist determined gnawing. *Mastomys* has destroyed aluminium cages, and steel or galvanized sheet are indicated. Water need not be provided for 48 hours so long as potatoes and carrots are available.

Handling

Initially *Mastomrs* was both wild and aggressive, and could be handled with forceps only. Attempts at checking escape by slamming down the lid led to many deaths from fractured necks. Much better results followed when the top and floor of the cage were made of mesh.

Table 5.1: Composition of Diet (per cent).

Straight run yellow mealie meal	32.7	Wheaten bran	9.1
Pure germ meal	17.0	Lucerne meal	4.5
Ground nut-cake meal	9.1	Fermentation by-products	0.9
Meat meal	4.5	Calcium	0.5
Fish meal	5.4	Fine salt	0.5
Oatmeal	11.3	Molasses and sugar	4.5

This permitted observation without lifting of the lid, and so avoided frightening noise and sudden light. It also reduced the frequency of cage-changing. The animals can now be handled without gloves, though they are not nearly as tame as laboratory rats.

On being handled, some adults, although not aggressive, make a soft high-pitched wheezing which continues for about a minute. Highly aggressive animals emit a highpitched shriek just before attempting to bite. Teeth chattering is also a common alarm reaction.

Nutrition

Compressed pellets form the standard diet. Their composition is given in Tables elsewhere in this chapter. One animal consumes approximately 6 g a day. An average analysis reveals 18.4 per cent of protein, 5.1 of fat, 5.8 of fibre, 8.0 of moisture, 6.4 of ash and 56.3 of available carbohydrate.

Coprophagy has been observed.

The Biological Laboratories at Pollards Wood of the Chester Beatty Research Institute maintain a colony of *Mastomys* on rat diet 86 supplemented with whole oats and brown bread given once a week.

Table 5.2: Amino-acid Additives (per cent).

Arginine	1.31	Lysine	0.84
Cystine	0-31	Methionine	0.36
Glutamic acid	2.10	Phenylalanine	0.99
Glycine	0-92	Treonine	0.78
Histidine	0-45	Tryptophan	0.27
Isoleucine	0.82	Tyrosine	0.78
Leucine	1.34	Valine	1.14

Water

Glass water-bottles are used; they are fitted with glass tubes of gauge 9 mm with their ends closed to a diameter of $^{7}/_{16}$ in (11 mm); these are held in stoppers made from short lengths of rubber tubing.

The insertion of glass tubes into the bottles is accompanied by some risk of fracture and the hand should be gloved. Stainless-steel tubing would not enable one to detect interruption of the flow by bubbles.

BREEDING

Physiology

In our colony, perforation of the vagina occurred at a medium age of 41 to 44 days, with a range of 38 to 47 days. The mean weight of females 42 days old is 20.2 g. Males at this age weigh 25.8 g.

The mean age at first litter (in 231 pairs) was 99.2 days. The distribution is markedly skewed, and with 5-day intervals the mode was 80 to 84 days, the youngest being at 69 days.

Since the gestation period is 23 days, the mean age at first conception was 76 days and the modal age 57 to 61. The mean age at last litter in the R line (81 pairs of the 11th to 20th generation) was 277.3 ± 7.5 days, the maximum age recorded being 401 days.

Davis (1963) has recorded a maximum age of 672 days, with mean of 7.6 litters per lifetime, and a mean of 7.4 young per litter. The lower rates in inbred colonies may reflect lack of hybrid vigour. The mean weight at 84 days is 48.3 g in males and 41.1 g in females. A weight of 40 g can be taken as a rough indication of sexual maturity.

The ventral prostate in the female consists of a pair of lobes, homologous with that in the male. It undergoes hypertrophy during the luteal phase of oestrus, and this condition persists through gestation.

Changes in the vaginal smear have been described by Johnston and Oliff (1954) who state that pro-oestrus and metoestrus both last 1.9 ± 0.15 days. Re-examination of these changes shows that the average cycle is 7 or 8 days. The cycle is as follows:

(1) Pro-oestrus: 1.9 ± 0.15 days. Abundant mucus streaks with nucleated squames and mucous cells;
(2) late pro-oestrus: large epithelial squames with degenerating nuclei, cornified squames, and mucous cells;
(3) oestrus: clumps of cornified, non-nucleated squames and mucous cells;
(4) early metoestrus: few leucocytes, and degenerating cornified squames;
(5) late metoestrus: abundant leucocytes, degenerating cornified epithelial cells and non-cornified cells and mucous cells;
(6) dioestrus: abundant leucocytes, nucleated squames, and mucous cells. No cornified cells.

Section of the vagina in both pro-oestrus and post-partum oestrus reveals a vaginal epithelium composed of a superficial mucous-cell layer, several times thicker than the underlying stratum germinativum. This is delaminated at or just before oestrus.

At oestrus the cornified layer is superficial; leucocytes accumulate beneath the epithelium in early metoestrus, and form pockets in the superficial epithelium in late metoestrus.

Post-partum oestrus occurs during lactation, 2.6 ± 0.52 days after parturition. The number of litters born varies with the stock. Hybrid vigour is striking. Litters up to 16 have been produced, but the average is about 8.

Management

During breeding the animals are given newspaper as nesting materials. Parturition is usually at night but also occurs during the day. Weaning takes place at 21 days, gestation is 23 days and post-partum ovulation occurs, so that in a good stock litters can be expected every 25 days.

At birth the young are covered with sparse hairs, and vibrissae are present. The eyes are closed, and in the R strain open at 15.2 days with a range of 13 to 17 days. As Meester also found, it may take one or two days for the eyes of all the young of a single litter to open.

Newborn males and females can be distinguished, as with rats, by the size of the genital papilla and its distance from the anus, this distance being more than 2 mm in males and less in females. An excellent description of their growth and development is given by Meester.

The pups squeak on the day of birth but are silent thereafter; they make smacking noises with their lips and also suck their feet.

Apart from the problem of litter-eating, poor breeding has been noted in brachyuric and umbrous lines, which died out in consequence, and Menzies (1957) has described genetic male sterility in the pale-grey variety.

He found spermatogenic arrest after the primary spermatocyte stage, and considered that the coat-colour and sterility were controlled by a single pair of genes. The pathological changes described are a non-specific testicular response.

In Sierra Leone, Brambell and Davis (1941) found a greater proportion of females pregnant in the wild during October and November than at other times, i.e. during the winter months, although they note that there is no obvious climatic reason why breeding should not be continuous.

Chapman *et al.* (1959) report that in Tanganyika breeding occurred from February to November with more young in April and May, i.e. mainly at the end of the rainy season and at the beginning of the dry season, when external conditions were most favourable.

Inbred Lines

In one inbred line (DB, a Baragwanath dilution) now at the 17th inbred generation. accidental placental haemorrhage has been a very common cause of death during or after parturition-not necessarily at the first litter.

Another strain, R, has been produced after 20 generations of inbreeding. This is a wild-type agouti, moderately susceptible to spontaneous carcinomas of the stomach, and very susceptible to papillomatosis of the skin.

Genetics

The South African species has 36 chromosomes, first demonstrated by Matthey (1954) (reported as *Mastomys coucha,* i.e. *natalensis)* and confirmed by Huang and Strong (1962). Different counts are reported from *Mastomys* from other regions. These represent distinct species, e.g. for *Mastomys* from the French Congo Matthey (1955) found 2n -= 32 and for *Mastomys coucha erythroleucus* (Temm) from the Ivory Coast he found a count of 40.

The marked differences in chromosomal length make the chromosomes easier to identify than those of the mouse. The X and Y chromosomes are among the largest; Matthey (1958) states that the Y is never smaller than the X, while Huang and Strong state that the X is the largest chromosome, and that the Y lies between autosomes 2 and 3.

The latter note that sex chromosomes showed no heteropycnosis in mitotic metaphase. Matthey (1958) concluded that the striking difference in karyotype supported the elevation of the subgenus *Mastomys* to generic status.

Certain distinctive coat-colours are available, but in the absence of the non-agouti gene they cannot be related to mouse coat-colour genes. The following types of pelt (with associated eye colours) have been distinguished so far:

(a) wild type, with black eyes, black ears, white paws, and dark tail;

(b) Baragwanath dilute (discovered in the original stock caught at Baragwanath near Johannesburg). This is a pale-bellied dilute, with dark ruby eyes and pale grey-brown ears and tail;

(c) Iscor dilution, obtained from specimens caught at Iscor near Pretoria. This gene is on a different chromosome from that of the Baragwanath dilution, but phenotypically the animals are similar;

(d) golden yellow, with brigot pink eyes, whitish yellow ears and tails, and whitish belly. This is an allele of the Ischr dilution. Heterozygotes of Iscor dilution and golden yellow are dilute with white ears, dark grey bellies, pink eyes, and white tails.

The genes for brachyury, umbrous coat and albinism have been encountered but have been lost.

Bleeding

Bleeding is achieved most simply from a nick in the tail or from cardiac puncture under anaesthesia. Tail veins are not convenient and attempts at collection of blood from the conjunctival sac have also been unsuccessful.

Anaesthesia and Euthanasia

Mastoazt's is very sénsitive to all common forms of anaesthesia, especially to pentobarbitone sodium, of which 8 mg per 100 g body-weight usually proves fatal. Ether is simple to use.

Diseases and Parasites

So far the only important problems have been the control of external parasites, diarrhoea, pneumonia, cannibalism, and the spread of papillomas.

Zumpt (1961) gives a full account of the parasites found on the species. Most of them appear to be of no importance in the laboratory.

6

NEWTS AND SALAMANDERS

In maintaining Urodela in the laboratory we may distinguish between (1) collecting the animals in nature and keeping them only temporarily, and (2) maintaining and breeding them permanently. In the first case the animals are usually collected during the breeding season, when they concentrate in large numbers in ponds and streams.

The husbandry then only serves to keep them in good condition for a certain period of time and, if necessary, to let them spawn. Usually such animals will not be kept longer than is necessary for experimental use.

When there is no convenient source from which they can be collected the best alternative is to maintain a colony of breeding adults. This, however, puts much higher demands on the care and accommodation required, since spawning will occur only when the animals are in healthy condition.

There are a few species of Urodela that are known at present to adapt themselves successfully to laboratory life. All these are species which can be adapted to a permanently aquatic mode of life, namely *Ambystoma mexicanum* (Mexican axolotl), *Pleurodeles waltlii* (Spanish salamander) and *Triturus pyrrhogaster* (Japanese newt).

In this chapter particular attention will be paid to the axolotl, which is the species most commonly used in the laboratory. In the discussion of other species reference will often be made to the information given for the axolotl. For a better understanding of the requirements for keeping these animals in good condition in the laboratory it has been thought advisable to include some information concerning their general biology and the conditions prevailing in the natural habitat of the species concerned.

THE MEXICAN AXOLOTL
(Ambystoma mexicanum)

Normal adult axolotls retain the external gills and thus do not change from an aquatic to a terrestrial phase of life. This makes them convenient as laboratory animals. They can be kept with rather simple accommodation and at little cost, but they require regular care and control.

They have a relatively high susceptibility to fungal infections. The species is widely used for experimental embryological research and for investigations on regeneration. The eggs can be obtained in large quantities over a long period of the year and are easy to decapsulate.

Ambystoma mexicanum belongs to the family Ambystomatidae. This family is characterized by the rather frequent occurrence of partial or complete neoteny; a condition in which sexual maturity is attained although several essentially larval features, such as external gills, tail-fin etc., are retained.

In the genus *Ambystoma,* Tihen (1958) places the axolotl in a distinct group, the members of which are characterized by being consistently neotenous under natural conditions. Some American authors even denominate this group as a separate genus, using the name *Siredon mexicanum.*

Geographic Distribution and Ecological Data

In nature the axolotl is found only in Mexico, in the Xochimilco and Chalco Lakes. Gadow (1903) describes the ecological conditions. The lakes are situated at a height of 2,250 m in a mountainous and fertile environment and get their main water supply from deep, clear springs.

The lakes are 1.50 to 3 m deep and have many small floating peat islands. In summer the water is covered with a layer of water plants. The aquatic fauna is very rich, with many small fishes, insect larvae, *Daphnia,* worms etc. Axolotls spawn in the beginning of February; the larvae grow very rapidly and reach adult size in June.

General Characteristics

External, Appearance and Morphology

The adult axolotl has a dark brownish-black skin with numerous black dots and a number of diffuse dark grey-brown spots. These spots occur particularly on the limbs and the tail. Old or not-entirely-healthy animals often show an increase in size of the greybrown spots, and generally become more greyish in colour.

The belly is dark purple-grey with black spots. Animals that have not yet attained sexual maturity have an olive-brown skin with distinct black dots. Axolotls may attain an ultimate length of nearly 30 cm. They have well-developed external gills, and a long tail with a dorsal and a ventral fin.

The males are more slender than the females, and generally have a longer tail; the margin of the cloaca is markedly swollen, particularly during the reproductive season. The females have a considerably stouter appearance and a flat cloacal region.

A description of axolotl anatomy and histology is given by Brunst (1955b).

Metamorphosis

Although the neotenic form prevails in nature as well as under standard laboratory conditions, it is known that the axolotl has not completely lost the ability to metamorphose. Metamorphosis may take place particularly in young axolotls about 15 cm in length either

(1) when they are fed with thyroid glands or organic iodine preparations, or injected with thyroidin or pituitary extracts; or

(2) when they are gradually adapted to a terrestrial mode of life. It has been observed thatt in some individuals metamorphosis occurs spontaneously under standard aquarium conditions. Metamorphosed young axolotls about 15 cm in length can continue aquatic life without any difficulty. Older metamorphosed animals, however, often drown if a terrestrial environment is not provided. Metamorphosed axolotls accustomed to terrestrial life temporarily resume an aquatic mode of life during the phase of reproduction.

After metamorphosis the axolotl has a smooth, dark-grey skin with a large number of distinct small yellow spots. Regular sloughing takes place, in contrast to the neotenic form. The gills disappear and the gill slits close. The tail-fin disappears and the tail assumes a roundish stout appearance similar to that in other land salamanders.

Regeneration

An interesting characteristic of axolotls is their high regenerative capacity. Regeneration is observed after amputation of limbs, tail and gills. In adult animals regeneration takes place at a slower rate than in young animals.

Genetics

There are two established races of axolotls: a black race and a white one. The colour of the wild stock is black; whiteness is caused by a single recessive Mendelian factor. The animals homozygous for the recessive gene vary from completely white to partially white with black or greyish patterning on the head and dorsal part of the body.

According to Woronzowa (1929) the skin of the white axolotl contains melanophores, though their number is considerably reduced, and moreover it has a reduced susceptibility to the melanophore-expanding action of the pituitary hormone.

Fertilization

As in all newts and salamanders, insemination is internal and takes place by means of spermatophores. In axolotls the spermatophore is deposited by the male during courtship; it consists of a cone of transparent jelly secreted by the male cloacal glands and bearing a white mass of spermatozoa at its apex.

When it is deposited, the base of the cone adheres to the substrate: the female immediately swims over it and takes up the spermatozoa into her cloaca. The sperm is stored in the spermathecal tubules of the cloaca until the eggs leave the oviduct and pass through the cloaca.

Before oviposition several spermatozoa have penetrated the jelly capsule of the egg. In all Urodeles several spermatozoa usually enter the egg; only one of them fuses with the female pronucleus, and the others degenerate.

Eggs

In the laboratory eggs can usually be obtained during eight months of the year, from November to June. One female lays some hundreds of eggs over two to three days. The eggs have a diameter of 2.2 mm and are surrounded by a capsule consisting of one elastic and several jelly layers.

Accommodation

Adult axolotis can be adequately kept in asbestos cement containers. This material is strong, durable and relatively cheap and does not give off toxic substances. Before use new containers must be continuously rinsed for three to four weeks, in order to remove excess calcium, and subsequently the inner surface should be coated completely with a

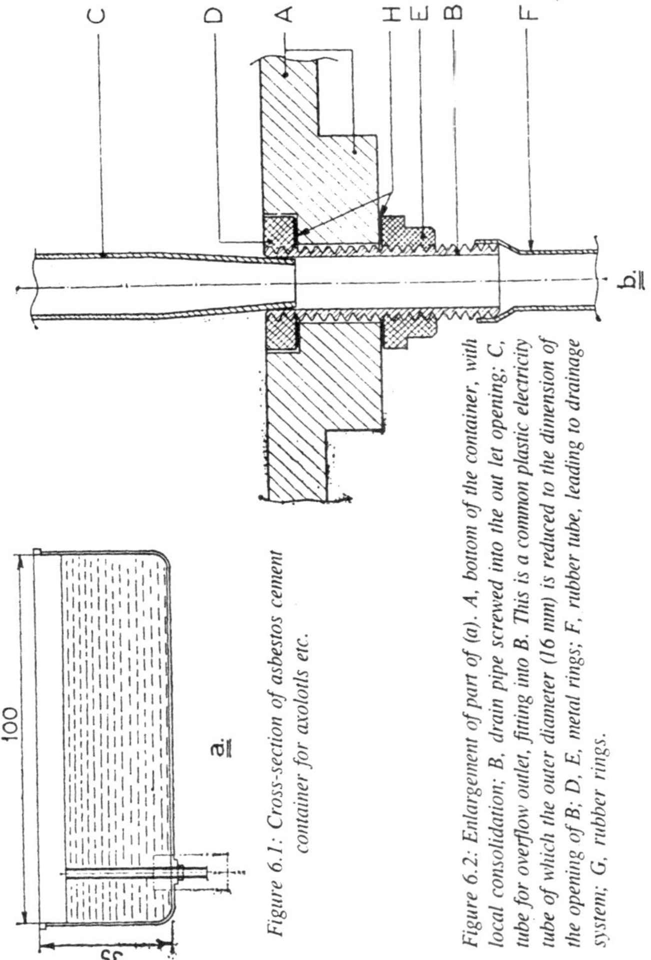

Figure 6.1: Cross-section of asbestos cement container for axolotls etc.

Figure 6.2: Enlargement of part of (a). A, bottom of the container, with local consolidation; B, drain pipe screwed into the out let opening; C, tube for overflow outlet, fitting into B. This is a common plastic electricity tube of which the outer diameter (16 mm) is reduced to the dimension of the opening of B; D, E, metal rings; F, rubber tube, leading to drainage system; G, rubber rings.

waterproof, chemically inert varnish. Aquaria constructed of glass with a metal framework and bottom are also useful, provided the metal parts are well varnished.

Other materials such as lead, zinc etc. are toxic to axolotls, and moreover have the disadvantage that the varnish comes off rather easily; hence containers made of such material have to be revarnished repeatedly.

Each container should be provided at the bottom with an outlet for complete drainage. In this opening a shorter or longer tube can be fitted to provide an overflow outlet.

In containers with a surface of 50 × 100 cm, 8 to 10 adult animals can be placed. If it is necessary to keep the animals separate, such containers can be divided into nine compartments of 16 × 32 cm by partitions of perforated waterproof hardboard, which are about 5 cm higher than the water level, and which have some larger openings at the level of the surface of the water for removal of the surface film.

It was found that the height of the water level has an influence on the condition of the gills of the axolotl. In shallow water the gills tend to degenerate, but a water level of about 25 cm keeps them in good condition. When the walls and partitions are at least 5 cm higher than the water level the axolotls will not jump out and it is not necessary to cover the containers.

No sand, gravel or aquatic plants are needed; this considerably facilitates the care of large numbers of animals. However, the presence of some algal growth on the bottom and side walls of the container is recommended.

Water

In many laboratories the quality of the tap water is not suitable for keeping the susceptible skin of axolotls in a healthy condition. The harmful influence can be reduced considerably if the water is allowed to stand exposed to the air for some days so that chlorine can disappear and oxygen be taken up.

Axolotls, like fishes, should be kept in relatively shallow water with a large surface area, so that adequate exchange of oxygen and carbon dioxide can take place. The gas exchange can be accelerated by the use of a mechanical water-aerator, which serves a threefold purpose:

(1) it causes turbulence of tile water, so that the water from deeper layers comes to the surface regularly;

(2) it removes the greater part of the surface film from the water;

(3) the small air bubbles offer a relatively large surface area to the water phase. The surface film is a thin bacterial layer, which is often formed on standing water and seriously interferes with gas exchange.

It can also be removed by a temporary slight raising of the water level, which causes the film to disappear through the overflow outlet.

Axolotls kept in well-aerated standing water do not suffer as much from skin infections as those kept in running tap water even when this is of a good quality. Apparently the axolotls form their own protective substances, which are constantly removed in a running water system. It is not harmful to supply fresh tap water two or three times a week in a quantity sufficient to replenish the amount removed during siphoning-off of detritus, and to let the surface film run off.

However, care has to be taken that not too large a part of tile total water volume of the container is renewed.

Light

Axolotls can be kept in ordinary daylight, or in artificial light which is switched off during the night. In the latter case the day period should be adapted to the day-length of the season.

The increase in day period in January apparently has a stimulating effect on reproduction. Another important function of proper illumination is to encourage algal growth on the walls of the container. This considerably promotes the clearness of the water and reduces the nitrate concentration.

A too abundant growth of algae, must, however, be prevented, since axolotls suffer very much from the nitrates released by putrefying algae. The entire layer of algae has to be removed now and then, especially at times of decreasing day-length and decreasing temperature.

Correct illumination is given by fluorescent lamps of white/pink colour, giving 40 per $2m^2$ of surface, and placed 40 to 50 cm above the water level.

Temperature

Axolotls live well at temperatures of 16 to 20°C (61 to 68°F). They should not be kept at temperatures higher than 22°C (72°F), but can tolerate lower temperatures; they thrive in open-air ponds even when these are covered with ice in winter.

NUTRITION

Axolotls feed very well on beef heart from which fat, tough fibres, etc. have been carefully removed, and to which are added 0.3 g of powdered chalk ($CaCO_3$) and 200 i.u. of each of vitamins A and D per 100 g. Either a fat-soluble or a water-soluble vitamin preparation may be used.

As long as the animals are not older than one and a half to two years, and as long as they are kept together in a tank, one can simply mince the meat, mix it thoroughly with chalk and vitamins, and put it into the container. The animals will seek it themselves.

Animals which are kept separately, however, do not seek the meat and have to be individually fed by hand. The same holds for older animals. The meat should be cut into strips of about 30 × 4 × 4 mm and then mixed with chalk and vitamins. It is important to offer strips of meat of a proper size.

If a piece is too large, it will not be readily swallowed and more food will be refused though the animal's appetite is not yet satisfied. If the strips are moved gently with a pair of forceps in front of the snout, the animals usually snap at it vigorously. Since they often snap also at the forceps one should use forceps with blunt tips.

They have to be fed two or three times a week. During spring and summer they usually eat three to four, in autumn only one to two strips each time. Usually females eat more than males. Individual strips have to be offered with suitable pauses, otherwise they will be regurgitated.

Instead of beef heart, beef or beef liver (no pig liver) can be given. The use of liver has the advantage that the addition of vitamins can be omitted, but it has the disadvantage that the water becomes rather dirty.

This is reduced when the cut strips are put into a dish, briefly flooded with water, and stirred, and the water is then poured off. The easiest way to prepare the strips is to cut the liver while it is still frozen.

A few hours after every feeding, droppings and remains of food should be siphoned out. If minced meat is given it is advisable to remove faecal matter before feeding, since the droppings, which are covered with a thin pellicle, may be broken by the active swimming movements of the animals.

BREEDING

Breeding Season

Sexual maturity in the male is indicated by a considerable swelling of the margin of the cloaca, caused by an increase of the size of the cloacal glands as they become functional. In the female a conspicuous roundish body shape indicates the presence of eggs in the body cavity.

External sex characters begin to appear in males at 7 months, whereas a female cannot be recognized with certainty before 9 or 10 months. Sexual maturity can be reached in 1 year, but the first batch of eggs is small. The best breeding age is 2 to 5 years.

The breeding season lasts from November until June. Males may be remated after 1 to 2 months, females after 2 to 3 months; consequently they can be used several times during each season. In order to keep axolotls in good breeding condition it is important to give them an opportunity to lay eggs at least once a year.

Induction of Breeding

Breeding can be induced by a sudden drop in temperature. The necessary difference in temperature depends on the sexual condition of the animals, the time of the year, etc., but usually a difference of 10°C (50°F) is sufficient. If the laboratory has tap water running at 10 to 12°C (50 to 53°F), the animals to be used for breeding can first be kept for 1 or 2 weeks at 20 to 22°C (68 to 72°F).

During this time feeding should go on as usual. Temperatures higher than 22°C (72°F) should be avoided, since they reduce the quality of the eggs. If one has no running tap water of a low temperature at one's disposal, crushed ice can be added.

All through the year males and females are kept in separate containers. Two days before the eggs are required one female and one or two males are put together in a clean, large container (bottom area 50 × 100 cm) with about 100 l of fresh water at a low temperature.

The bottom of this container should be covered with clean slates, and between the side walls a number of strong, flexible twigs for oviposition should be fitted at about half of the height of the water level.

The container has to be sheltered from light, since sudden movements made in its neighbourhood or the switching on of a light can easily disturb courtship.

The low temperature should be maintained at least until the time of spermatophore deposition. Usually one finds spermatophores, deposited on slates or twigs, one day after the temperature shock. Sometimes one male deposits a large number of spermatophores (up to 20), but it has been found that even when only one spermatophore is present, this can be sufficient to fertilize a large batch of eggs.

When deposition of spermatophores is delayed, it may still be induced by raising the water temperature after a few days. Apparently low temperature partly inhibits the mechanism of spermatophore formation, and a rise of temperature causes a release.

The animals should not be fed from the day on which they are put into cold water until the end of oviposition. For every planned spawning at least two couples should be used.

Oviposition can be stimulated by injection of gonadotropin or by subcutaneous implantattion of frog pituitaries. However, in axolotls the eggs obtained are unfertilized. Brunst (1955a) and Humphrey (1962) describe a method of artificial fertilization of eggs, dissected from the oviducts.

SPAWNING AND COLLECTION OF EGGS

One day after spermatophore-formation the female begins to deposit her eggs on the twigs. For some time before oviposition she swims in a remarkable floating manner.

It is important that the animals shall not be disturbed during the early phases of oviposition, since this may stop the spawning process. As soon as about a hundred eggs have been laid, the twigs can be taken carefully out of the water.

The eggs in their jelly capsules can easily be removed with a pair of forceps with curved tips. The branches may be put back into the container after having been well scrubbed. The duration of oviposition is partly dependent on the temperature.

For experiments in which early stages are required the temperature should be kept at 12°C (53°F) throughout oviposition. At this temperature oviposition takes about three days. Not only the slower development of the eggs, but also the availability of freshly-laid eggs over a longer period can bean advantage for experimental embryological work.

A temperature lower than 12°C(54°F), though not harmful to the process of spermatophore formation, often blocks oviposition. The number of eggs varies with the season and with the age of the female, and may range between 300 and 2,000 eggs per spawning. The usual number is 500 to 700.

Rearing of Eggs

Failures in rearing eggs and larvae are usually caused by fouling of egg capsules and culture media. Hygienic conditions are therefore essential. The material must be inspected regularly and dirty dishes thoroughly cleaned.

Space and accommodation

The water in the dishes should not be deeper than 2.5 cm; the eggs have to be placed in a single layer and should not be packed closely. In order to keep out dust and to prevent rapid evaporation of the water the dishes should be almost completely covered with glass plates. Ready gas-exchange, however, must not be prevented.

Temperature

Eggs can be reared at room temperature, the optimal temperature for development being 14 to 18°C (57 to 64°F), although they can stand temperatures from 8 to 22°C (46 to 72°F). Sudden temperature-changes must be avoided.

Therefore the eggs that have been removed from the twigs are put in large glass dishes with water of about the same temperature as that in the container used for oviposition.

Light

There are no special light conditions, except that the eggs should be protected from direct sunlight.

Cleaning of the egg capsules

The surface of the jelly capsules is a favourable place for bacterial growth; hence the outer layer of the jelly capsule, which swells considerably by water uptake, has to be removed after some days. The simplest way to do this is to place the eggs on the palm of the hand and to remove the outer, softer jelly from the deeper, firmer layer with a blunt knife. It may be useful afterwards to dip the eggs into 70-per-cent ethanol for about 10 seconds for partial sterilization of the jelly surface.

Rearing of Larvae

Space

After about two weeks (at 18°C, 64°F) the larvae slip out of their capsules. Empty capsules are at once removed and the newly-hatched larvae are transferred in a pipette with rubber bulb to glass dishes 20 cm in diameter. In order to keep the dishes as clean as possible no sand or water plants should be added.

During culturing, overcrowding has to be avoided. The more space the larvae have the more quickly they grow. In crowded conditions, moreover, the larvae often attack one another. This leads usually to death, since wound-healing is very slow.

Animals which once have a wound will be attacked repeatedly by the others. The surface areas and depths of water generally required for successive stages of axolotl larvae are given in Table elsewhere in this chapter.

These data hold for animals which are kept together and when they are fed abundantly. There is always a variation in growth-rate of larvae kept in the same container. When the differences become large, the bigger animals tend to attack the smaller ones.

Therefore now and then the larvae have to be sorted into groups of larvae of similar size. The larvae should be distributed in several dishes in order to localize possible diseases, which usually spread rapidly among all larvae in the same dish.

Table 6.1: Spece requirements for growing larvae of ambnystoma mexicanum.

*Approximate length (cm)**	*Average area of bottom surface per animal (cm²)**	*Depth of water required (cm)*
1.3 (2 to 3 weeks, just hatched)	10	3
3 to 4	40	3
5	100	5
10 (4 months)	250	15
15 (6 months)	500	25
18 (7 months)	600	25
25 to 30 (18 to 24 months and older; adult)	500-600	25

* The time required to reach this length at about 18°C (64°F) is given in brackets.

Light

Light conditions are similar to those for eggs.

Water

It is important, particularly during early larval growth, to use clear pond water, warmed up to room temperature. Water from aquaria which are in stable biological equilibrium can also be used.

These aquaria, however, should not have been inhabited previously by Anurans. These precautions with respect to the water in which the larvae are reared are necessary since the skin of urodele larvae is very delicate.

Common tap water usually has too low a salt-concentration or contains chlorine; it causes a bad condition of the larval skin and therefore promotes the chances of infection. Pond water or old aquarium water, however, apparently gives young larvae very good protection against diseases.

Breder (1931) found that the advantage of old aquarium water is partly due to its larger buffering capacity, but the main factor is the presence of a bacteria-bacteriophage equilibrium, which prevents fouling of the water. Larvae of about 5 cm can be reared further in tap water without danger.

Feeding

In general it is necessary to feed growing larvae abundantly. The best results are obtained when food is constantly present. The food therefore has to be supplied regularly, but no more food should be provided than the animals can eat in a day, in order to avoid fouling water.

The amounts have to be determined by experience. Two to three days after hatching the larvae start feeding on small moving aquatic animals such as *Daphnia* or larvae of *Artemia saliva*. Eggs of *Artemia,* obtainable from aquarium-supply houses, hatch in 2-per-cent NaCl solution at 28°C (82°F) after 24 hours and can be supplied after being rinsed with tap water.

These larvae can be used as a first food after hatching. *Daphnia* is an excellent food for amphibian larvae. Axolotl larvae grow very well and rapidly on a diet consisting of *Daphnia* only. This can be given until they have a length of about 10 cm.

Instead of *Daphnia* other small aquatic animals can be provided, such as other crustaceans and mosquito larvae, but large quantities of *Cyclops* should be avoided since this crustacean attacks the larvae and a healthy culture of young larvae can be ruined by them in one hour.

Geyer (1957) gives methods for culturing *Daphnia* and *Artemia* indoors. If it is difficult to obtain the large quantities of *Daphnia* required for feeding more advanced larvae, the diet for larvae of about 3 cm and more can be alternated with *Tuhifex* and *Enchvtraeus*. In the beginning only small worms or parts of worms need be fed.

Larvae of about 5 cm can also be fed with very finely minced meat, but this must be alternated with *Daphnia,* worms etc. A diet of exclusively *Tuhifex* and *Enchytraeus* does not give satisfactory results. Hutchinson and Hewitt (1935) maintain that this diet causes the most rapid growth, but it has also been observed that the larvae afterwards show defects.

Witschi (1938) advises feeding them from the beginning with small pieces of liver, offered individually on a dissection needle. This method offers the advantage of an easy control of the amount of food taken, and the water does not become dirty by spoilt remains of food.

The method will certainly be suitable in several experimental conditions, including the feeding of operated animals; for culturing larger groups of animals it is too time-consuming.

Hutchinson and Hewitt found that a diet of liver, in comparison with diets of *Daphnia, Tuhifex* and *Enchytraeus,* gives a considerably slower growth.

When the larvae are 10 cm long they can be gradually switched over to the diet for adult animals. However, they have to be fed every day until they have reached a length of about 20 cm, otherwise they will attack one another.

Animals of 18 to 19 cm are especially aggressive. It is for this reason that they need relatively much space, Table 1. The animals can be kept in separate containers during this period, but this has the disadvantage that the animals then have to be fed by hand; moreover the cleaning of separate containers takes more time.

Cleaning

At the end of every day detritus has to be removed. In the small glass dishes this can easily be done with a glass pipette with a rubber bulb. In the larger tanks, containing more water, the detritus can be siphoned out with a wide rubber or plastic tube.

The water removed has to be replenished with pond water at room temperature in the case of very young larvae, and with common tap water for older larvae. Whenever the water becomes cloudy or foul-smelling, the larvae have to be transferred immediately to another clean dish with fresh water.

ANAESTHESIA AND EUTHANASIA

Anaesthesia

The most satisfactory narcotic for larvae is MS 222 (tricaine methanesulphonate) from Sandoz Chemical Works, Basel, Switzerland. It is soluble in water. Copenhaver (1939) found that it has only a slight effect on heart-rate. It is less toxic and gives a more rapid recovery than chlorbutol (syn. chlorbutanol, chloretone).

The recommended concentrations for axolotls vary from 1:5,000 to 1:1,000 for larvae up to a length of 11 cm. The animals should be immobile and numb in about five minutes. Recovery takes place in 15 to 30 minutes in clean water.

Schotte and Butler (1941) report that larvae after a short narcosis in 1:5,000 MS 222 can be kept without ill effects for more than two consecutive days under narcosis in a physiological salt concentration with 1:10,000 MS 222. No cumulative effects have been observed. See also Staphanova *et al.* (1962).

For adult axolotis Stephenson (1947) advises a 1-per-cent solution of urethane (ethyl carbamate). It works quickly and seems to give no ill effects.

Euthanasia

The animal should first be immobilized in a 1-2-per-cent urethane or a 0.1-per-cent MS 222 solution and then be removed to a tin containing cotton-wool soaked in chloroform. For some purposes it can be preferable to kill the animal by placing it in a 5-per-cent urethane solution or by decapitation with an extremely sharp pair of scissors.

TRANSPORT

Within The Laboratory

Adult axolotls

Adult axolotls which have to be removed to another container should never be lifted with a net. This startles them to such an extent that they vigorously beat their tails and may damage these considerably. A suitable method for holding an axolotl.

After one hand has been moved gently towards the animal's head, the trunk is firmly grasped from above between the thumb and the forefinger and middle finger, just behind the forelimbs. At the same time the two remaining fingers enclose the head, so that the axolotl cannot slip away forwards.

One has to take care that the gills are not squeezed. The grip behind the anterior limbs also serves to reduce gill damage. The forelimbs come between the middle and third fingers. Simultaneously the other hand grasps the caudal trunk and anterior tail region from above in such a way that the limbs come between the middle and third fingers.

In this position the axolotl can easily be taken out of the water without slipping away and without vigorous movements of its tail. In the new container it should not be released before it is entirely under water.

Young axolotls

Young axolotls (10 to 20 cm) can be held with one hand, the head and the anterior trunk being grasped from above in the way described for adults.

Larvae

Very young larvae have to be transported from one container to the other with a glass pipette with a wide mouth and a rubber bulb. Older ones can be removed with a net made from a loop of iron wire with a piece of nylon net stretched flatly across it.

By Road, Rail, Air etc.

Preparations

Every journey over a long distance, particularly to foreign countries, must be prepared carefully in order to avoid needless delay on the way. The transport company involved should be consulted in advance with respect to:

(1) the most suitable route and timeschedule;
(2) the availability of an air-conditioned luggage cabin, so that excessive heat or cold can be avoided; and
(3) packing conditions. Usually the shipments have to be paid for in advance and cannot be insured.

For transport abroad one has moreover to be well informed about the requirements for import permits or health certificates. In general one should avoid sending live animals over the week-end.

At the receiving end arrangements have to be made to collect the consignment as quickly as possible; hence the package should be provided not only with the full address but also with the telephone number of the consignee and the transport company should be requested to give notice to the consignee immediately on arrival of the animals.

Moreover the consignee has to be informed in advance about the details of the transport and the planned hour of arrival. He may be advised to contact the receiving office in advance in order to arrange for collection.

Packing

Axolotls can be transported in two firm plastic sacs (one inside the other), filled with so much water that the water level is 2 or 3 cm above the animals, and with at least twice this volume of oxygen. The sacs can be tied up with a rope and should be placed in a corrugated cardboard box.

During transport axolotls can be packed rather closely without danger of their attacking one another; e.g. six adult animals in 3 l of water. To minimize water fouling, they should not be fed on the day before transport nor during the journey. No water plants should be added.

DISEASES

Skin Disease

In this section only a few common diseases occurring particularly in axolotls will be mentioned. A general review of diseases of Amphibia is given by Reichenbach-Klinke and Elkan (1965).

Fungal Infections

Axolotl larvae are very susceptible to fungal infections with small tufts of whitish threads visibly protruding from small openings in the skin. These appear initially at the base of the tail and on the gills and head, but spread very rapidly all over the body and cause death within two days.

This infection may occur when the larvae are cultured in dusty and dirty rooms, when they are placed in the neighbourhood of stables, when dead organic matter accumulates in the dishes, and so on.

Treatment comprises placing the larvae in cleaner surroundings and in a fresh solution of 1:100,000 chloraminium (para-toluol sodium sulphonchloramide), added in powder from to pond water or water from aquaria in good biological equilibrium.

Chloraminum is a strong disinfectant, used against several fish diseases. Its chief action is due to oxygen liberated initially; the liberated chlorine is immediately bound as NaCl, so that the animals can stay in the solution and can be fed normally.

One day after treatment the mould threads will have disappeared completely and on the formerly infected spots only a skin-thickening or a haemorrhage will be visible.

These disappear rapidly. No abnormalities in growth and development have been observed. In severe cases a new dose of chloraminum may be added after 3 to 7 days (0.5 to 1 g per 100 1). However, one has to be careful with chloraminum treatment on account of weakening of the skin which sometimes occurs and makes the animals extra-susceptible to infections.

Another drug, used against fungal as well as non-fungal (see below) skin diseases, is mercurochrome. The animals are first put for 3 days in a concentration of 1:250,000, then for one week in pond water, and finally for another 3 days in a fresh mercurochrome solution of the same concentration as before.

In order to treat mould infections Detwiler and McKennon (1929) kept larvae older than stage 37 Harrison continuously in a mercurochrome solution of 1:750,000 in spring water over a period of 2 months without noticeable effect on growth.

Non-fungal Skin Disease

An occasional skin disease in adult axolotls is an infection which causes the formation of large whitish spots, which appear at first in the border region of trunk and tail and on the gills, lips and legs. It causes a reduction of the gill filaments.

In more severe cases the white spots are also formed on the tail and near the dorsal fin of the trunk. A slight touch of this disease is not harmful, until the damage to the gills finally results in a poor general condition.

A severe attack of the disease can be fatal. Nothing is known of its cause except that it is not a fungus. It appears particularly when the animals are kept in constantly running tap water. Apparently this washes out some property of resistance from the skin.

It also occurs, however, in other circumstances. For treatment a temporary rise of the water temperature to 20 to 22°C (68 to 71°F) or the mercurochrome treatment mentioned above may be tried.

Malnutrition

Brunst (1955b) reviews various tumours, tumour-like structures and abnormal pigment spots observed in axolotls and describes malformations such as sexual underdevelopment due to insufficient feeding during larval development and the hereditary absence of one or two forelimbs.

Young axolotl larvae which are not fed properly sometimes swallow air. Air bubbles accumulate in their stomachs so that they float on the surface and cannot feed. It is very difficult to get rid of these air bubbles.

Hamburger (1960) advises putting the animals in shallow water with plenty of food, or removing the air by puncturing the stomach with a glass needle, or squeezing out the air through the mouth, holding the animal with one pair of forceps and pinching behind the air bubble with another pair.

When the larvae have been cultured on sand the accumulation of air bubbles is fatal, since then usually also sand accumulates in the stomach. This cannot be removed.

First Aid

For the treatment of wounds the animal must be removed from the water and while it is held firmly the wound must be painted with a 2-per-cent mercurochrome solution. Heavily damaged legs must be amputated and then painted with mercurochrome.

Brunst (1955a) observed that all types of healing processes occur more rapidly at a low temperature. After operations the animals were kept several days in a refrigerator just above freezing, without food. The healing process occurred quickly and without infection.

When an animal in a colony shows symptoms of a disease it should be isolated at once. It is advisable to kill it, but if an attempt is made to save it, precautions must be taken to ensure that the infection does not spread in spite of isolation. Separate pipettes, siphons etc. should be used for the infected animals, and the person who attends to them should wash his hands afterwards.

It is very difficult to maintain complete isolation between separate containers in the same aquarium room, and it is preferable to kill sick animals, and also those which are not entirely healthy since they have a decreased resistance against diseases.

Dead animals should be removed immediately. When dissection does not give a clear indication that death has a non-infectious origin, the container in which a dead animal was found must be cleaned and sterilized thoroughly. Other animals that were in the same container should be treated with chloraminum or mercurochrome, and afterwards watched closely.

When many dead individuals are found in the same container within a short time it is advisable to kill all the remaining animals, and sterilize the container and associated apparatus thoroughly, in order to prevent the disease from spreading to other groups.

OTHER AMBYSTOMA SPECIES

The species *A. punctatum, A. tigrinum, A. opacum* and *A. jeffersonianum* are widely used in the U.S.A. for experimental embryological research. The breeding season varies considerably with the latitude of the places of origin.

Therefore in the U.S.A. many laboratories have developed an extensive system of mutual supply of material, according to which either just-spawned eggs or mature females are transported from the places of origin to other parts of the country.

Data on range, breeding season etc. are given by Bishop (1943). Hamburger (1960) and Rugh (1962) give more details about larval development. The methods of rearing and feeding larvae do not differ in principle from the methods described for the Mexican axolotl.

Only *A. tigrinum* seems to be amenable to permanent laboratory conditions. Hutchinson (1950) kept this species for three consecutive seasons under special conditions in which even spawning took place spontaneously.

THE SPANISH NEWT

(Pleurodeles Waltlii (Michah)

This species can easily be kept in the laboratory, under conditions in which it is not allowed to come on land. It is rather resistant to infections and can live for many (up to 20) years. It is very useful for experimental embryological research since the eggs, which are easy to decapsulate, can be obtained in large quantities over a long period of the year.

Geographical Distribution and Ecological Data

The species is found in the southern half of the Iberian peninsula and in Morocco, where it lives in small ponds with a considerable amount of vegetation. It spends most of its time in water; when their ponds dry up the animals aestivate under stones etc..

General Characteristics

The genus *Pleurodeles* belongs to the family *Salamandridae*. It is distinguished from the genus *Triturus* by having long sharply pointed ribs, which end in skin thickenings on the flanks. In *P. traltlii* these thickenings have a yellow-orange colour.

The ribs may perforate the skin and remain exposed without detriment to the animal. They are believed to serve as a protective weapon. Pleurodeles also differs from *Triturus* in having an amplexus, and not a courtship dance, before the spermatophores are deposited.

Normal adults have a rough yellow-grey skin with many dark spots. The usual length attained in the laboratory is 20 to 25 cm. Males are more slender than females and their forelimbs, which are longer and more sturdy than those of females, have a callosity at the inner side of the upper arm.

In males the cloaca is slightly more swollen than in females, the tail is longer and more fin-like, and it has a slightly orange lower edge. After metamorphosis the animals slough regularly and, like other newts, they eat their own old skin.

The female lays 200 to 800 eggs over two to three days; the eggs have a diameter of 1-7 mm and are surrounded by a soft jelly capsule.

Accommodation

A dozen adult animals are kept in containers measuring 50 > 50 cm with a water depth of 25 cm. No sand, gravel or aquatic plants are needed. A few rough stones, such as lava stones, or rough branches are placed in the water in order to help the animals to get rid of the sloughed skins.

The containers should be provided with an overflow outlet. They need not be covered. Although the animals like to go out on land, they should not be given the opportunity to do so. When on land they are difficult to feed and often stop feeding for a long period.

This causes considerable weakening and moreover increases the chances of infection. Adult animals can be kept in standing or slowly running tap water at 16 to 20°C (61 to 68°F), which should be well aerated. The container should be put in a sunny place.

Nutrition

Adult animals feed very easily on minced beef heart, which may be mixed with chalk and vitamins as for axolotls. The meat should be provided twice a week in a sufficient quantity, which has to be determined by experience.

It can simply be put in the container, Hamerton (1957) reports daily feeding with blowfly larvae. A few hours after feeding, the remains of food, droppings, and pieces of sloughed skin must be siphoned off and the stones or branches thoroughly brushed. Once or twice a year the containers must be emptied and cleaned with warm water.

Breeding

Sexual Maturity, Breeding Season

In the male, sexual activity is indicated by blackening of the inner side of the forelimbs, hands and fingers. The margins of the cloaca are slightly enlarged. I n the female a thickening of the anterior body region indicates the presence of many eggs in the body cavity.

External sex characters become visible at the age of 11 to 12 months, when the animals have a length of 11 to 15 cm. The first batch of eggs can be expected after about 16 months, when they have a length of 15 to 16 cm.

Gallien advises inducing oviposition each year several times, in order to keep the reproductive organs in good functional condition. The breeding season seems to vary in different stocks. Gallien (1952) mentions a period of September to May, Hamerton (1957) obtained eggs through the year, and from some stocks eggs can be obtained only from summer up to late autumn.

Spawning

The two sexes are kept in separate containers throughout the year. For spawning, one female and one male are put together in a rather small container, with water of the usual temperature (16 to 20°C, 61 to 68°F). No special light conditions are necessary.

The water should be well aerated and provided with twigs and slates for oviposition. Amplexus begins almost at once. The male approaches the female from below and clasps his forelimbs around the forelimbs of the female.

The two animals, the male below the female, swim together for several hours. The formation of spermatophores and the subsequent acceptance of the spermatophore can only rarely be observed.

At the beginning of the breeding season Gallien could observe several spermatophores, but later in the season they were seldom found. After 1 to 3 days the female begins to lay the eggs on the twigs and slates. The eggs are laid in small packets of about 12, with intervals of 10 to 30 minutes, so that it is easy to obtain eggs which are laid at the same time and of which one knows the exact time of deposition.

Oviposition continues for 2 or 3 days. The first batch consists of only about 150 eggs. Later the number increases from 400 to 800. It has been observed that a female which was kept isolated after oviposition subsequently laid a further 250 fertilized eggs after 2 months.

This indicates that many more spermatozoa than necessary for the fertilization of one batch can be taken up in the spermathecal tubes and that they remain fertile for several months.

Spawning can be reliably stimulated by putting the couple in a container placed in direct sunlight and filled with fresh water with several *Elodea* plants.

Rearing of Eggs and Larvae

In general the directions given for the axolotl can be followed. The eggs should not be placed at temperatures lower than 10°C (50°F). Although adult animals live very well in tap water, larvae should be cultured in pond water.

According to Gallien and Durocher (1957) at a temperature of 18°C (64°F) the larvae slip out of their capsules after 12 days (at about 11 mm); they begin feeding at about 17 days (about 12 mm) and metamorphose at 90 to 110 days (about 44 to 72 mm). Metamorphosis can be retarded by placing a small number of animals in a large volume of water.

Anaesthesia

For adult animals: 1 to 2-per-cent urethane (ethyl carbamate) or 0.05-per-cent chloretone is adequate; complete anaesthesia requires 15 to 30 minutes, recovery several hours.

In addition to MS 222 already mentioned, larvae can be anaesthetized with chlorbutol in a concentration of 0.005 per cent. About 10 minutes are required for complete anaesthesia and 15 minutes for complete recovery.

Transport

The animals can be transported in plastic or wooden boxes with adequate ventilation holes, covered by fine-mesh wire gauze positioned so that they cannot damage themselves by rubbing against the gauze. Some damp moss or fragments of wet polyester sponge are added to keep the atmosphere humid.

The use of wood fibre should be avoided since its sharp edges may damage the animals. They need not be fed during the journey. See also under axolotl.

Diseases

Diseases are seldom a problem in this species. A skin infection has been successfully treated with mercurochrome and chloramine solutions (see axolotl). In some cultures of young animals drowned animals are regularly found.

In most cases this is due to the animals not being able to get rid of their sloughed skins. Lava stones or rough branches, to which loose pieces of skin will stick while the animals swim, should be placed in the containers.

THE JAPANESE NEWT

Triturus pyrrhogaster (Boie)

Triturus pyrrhogaster adapts very well to laboratory conditions. It can easily be kept without special accommodation and with simple methods of feeding. A healthy stock, when given periods of hibernation, produces regularly fertilized eggs.

Egg-production can be increased by pituitary implantation. The species is much used for experimental embryological work, particularly in Japan. Decapsulation of the eggs, however, requires considerable skill.

Geographical Distribution and Ecological Data

T. pyrrhogaster occurs in Japan, where it is found in rice fields, ponds, ditches etc. with clear standing or slowly running water. It always has an aquatic mode of life except during the period of hibernation, when it hides in wet places located in crevices, under stones, fallen leaves etc.

General Characteristics

Adult animals are dark brown or black, with a red-orange belly beautifully patterned with black spots. The pattern varies considerably in individual animals. Characteristic of the species is also the marked protrusion of the parotids. The usual length of the animals is 12 to 17 cm.

Eggs are laid singly on water plants and, as in other *Triturus* species, the female wraps a leaf around each egg. The size of the egg is 2 mm. It is surrounded by a strong oval jelly capsule with a sticky outer layer.

Accommodation

Aquatic period

Males and females are kept together in containers measuring 50 x 100 cm with a water depth of 25 to 30 cm. No sand or gravel should be given; one or two flowerpots with water plants, viz. *Vallisneria* spp., are placed in the container.

The leaves of the plants should not be in contact with the side walls, so that the animals do not gain a support for climbing out. The containers need not be covered. It is advisable to aerate the water with a mechanical aerator and to keep the temperature at 16 to 20°C (61 to 68°F).

Hibernation period

In winter the animals are kept for 6 to 8 weeks without food at a temperature of about 3°C (37°F). A convenient place is the vegetable compartment of a refrigerator. Only some very damp moss or fragments of wet sponge need be added.

The hibernation compartment should be inspected every week, care being taken that the animals are disturbed as little as possible. The moss or sponge must be moistened again and dead animals removed.

Only healthy and well-fed animals can bear a hibernation period. They should be left unfed for two days beforehand to prevent intestinal disorders from undigested food.

Nutrition

This is the same as for the Spanish newt.

Breeding

Sexual Maturity, Breeding Season

Sexual maturity is reached after two years, but a large quantity of eggs can be expected only after three years. Sexual activity in the male is indicated by a greyish violet colour of the skin in the regions of the parotid glands, the flanks, and the tail.

The digits become elongated and the tail becomes more pointed and longer than usual. The skin is smooth and the margin of the cloacae onsiderably swollen.

In the female there are no external sexual characteristics except a swelling of the trunk region and a slight increase in the size of the cloaca. As usual in *Triturus* species the male deposits spermatophores after a courtship dance.

In nature eggs are laid from April to July. The period of egg-production can be regulated in the laboratory in relation to the period of hibernation.

Induced Spawning

From a healthy stock in which males and females are kept together, and which have been kept in temporary hibernation, one can usually obtain some fertilized eggs; this shows that males can produce spermatophores in captivity.

However, one can rely neither on the time at which the eggs are laid, nor on their number. Egg-production can be accelerated and controlled by implanting into the female small grafts of fresh pituitary glands from cows, female frogs, newts etc.

The best results are obtained by implantation, after anaesthesia, of a fragment of about 3 mm^3 in a pocket made underneath the skin of the lateral trunk region or the lower jaw. If too large a piece is inserted the female may lay a number of abnormal eggs.

The animal begins to lay eggs after a few days, and will regularly produce 4 to 5 eggs a day over a period of 1 to 3 weeks. When implantation does not lead to eggproduction, or when egg-production stops, another piece may be implanted. So one can work with one female for 3 to 4 weeks. After a rest period of two months (with normal feeding) new implantations can be made.

The optimum temperature for oviposition is 18 to 20°C (64 to 68°F). The female lays eggs easily on leaves of *Vallisneria.* Pituitary stimulation has not yet been successful in males that do not deposit spermatophores spontaneously.

Animals may be kept temporarily in an inactive condition at 10°C (50°F); it is then necessary to feed them. In order to get eggs in autumn the animals should be stored in hibernation until about one month before the eggs are required.

Rearing of Eggs and Larvae

In general the method as described for the Mexican axolotl can be used. The eggs can easily be removed from the leaves; it is not necessary to remove the outer jelly layers.

Hatching takes place when the larvae have a length of about 10 mm (about 4 weeks). They start feeding at a length of 11 mm, and metamorphosis takes place at a length of about 25 mm. The larvae can remain continuously in water in which some water plants are placed during and after metamorphosis, so that they can easily reach the water surface.

After metamorphosis, when they usually stop feeding on *Daphnia,* they can be fed with small worms until, at a length of about 5 cm (10 months) the diet can gradually be switched to very finely minced meat.

Anaesthesia

The following solutions are used; for adults, 1:1,000 to 1:3,000 tricaine methanesulphonate (MS 222 Sandoz), or 2-per-cent ethyl carbamate (urethane); for larvae, l-percent urethane or 0.05-per-cent chloretone.

Diseases

A number of symptoms, indicated in the German literature as *Molchpest,* are characterized by skin affections such as disturbances of sloughing, blisters at the snout, mouldy spots, haemorrhages and wounds which can be particularly serious on the legs and tail. Decreased appetite and a tendency to leave the water have been observed.

The cause is often unknown, but the disease *occurs* more frequently when the animals have been kept temporarily under less hygienic conditions, e.g. during transport or hibernation.

This disease is usually fatal. Chinosol has been recommended for immediate disinfection of the animals and their containers. Various other diseases occurring in Urodeles are described by Jacob (1909).

EUROPEAN *TRITURUS* SPECIES

The European newts are described by Gadow (1901) and Boulenger (1910). They live in water during the breeding seasons and on land during the remaining part of the year. When collected during the breeding season *T. alpestris* (alpine newt), *T. vulgaris* (smooth newt) and *T. helveticus (-T. palmatus,* the palmate newt) lay eggs in the laboratory for 3 to 4 weeks, *T. cristatus* only for a few days.

T. taeniatus, common throughout most of Europe, has a length of about 7 cm and the diameter of the egg is 1.25 mm. *T. alpestrisr* occurs mainly in mountainous regions up to 2,600 m in central Europe; it has a length of 7 to 10 cm and the diameter of the egg is 1.5 mm.

T. helveticus occurs in western central Europe up to 1,000 m and is particularly common in France; the length is 6 to 8 cm; the diameter of the eggs is 1.5 mm. Eggs are laid singly within folded leaves of water plants. The solid jelly capsule can easily be removed.

When breeding activity decreases the animals tend to leave the water. In aquaria they begin to swim restlessly along the walls, and try to climb out. Little is known about the conditions necessary for their terrestrial period of life.

Therefore it is preferable to return them after use to their natural habitat. According to Gallien and Bidaud (1959) *T. helveticus* can be kept permanently in the laboratory in water. In this condition even spontaneous reproduction can take place from December until May or June.

In *Triturus* neoteny occurs occasionally. The limbs, tail and gills have a high regenerative capacity.

Accommodation

About thirty animals can be kept in large containers of 50 *x* 100 cm, with standing or slowly running tap water at a level of about 20 cm. When spawning is desired there should be a ratio of one male to not more than three females.

Couples kept in individual containers need at least 6 1 of water. No light should enter the containers through the side walls, since this disturbs the animals. The bottom of the containers should be provided with a 3-cm-thick layer of thoroughly washed river sand in which a number of plants are fixed.

Convenient species for oviposition are e.g. *Myosotis aquatica, Mentha aquatica, Lpsimachia nummularia* and *Taraxacum officinale.* The containers must be well closed with iron or plastic gauze, which should not prevent the ready transmittance of light and fresh air. The temperature should be 16 to 22°C (61 to 71°F).

Nutrition

The better the animals are fed, the longer they will lay eggs and the more eggs will be produced each day. The best food for a larger group of animals is finely minced beef heart, prepared as described for axolotls, and provided three times a week.

The animals will seek it themselves. However, animals kept individually do not seek the meat. They have to be fed by hand or they should be given moving food such as earthworms; smaller worms can be given whole, larger ones should be cut in two or three pieces.

Some varieties of worms are not readily accepted by the newts. These should not be given. At the end of every day remains of food, decaying leaves etc., should be removed.

Breeding

Sexual activity, Breeding season

Sexual activity is indicated by a considerably intensified coloration of the skin. In T. *alpestris* the males become a violet-blue colour, with a light band with black spots on the flanks. The belly side is orange-red.

On the back a low crest is formed, with alternating blocks of black and yellow. Also the head, forelimbs and hindlimbs, and cloaca show black spots. The female is grey-brown on the back and flanks, with a less pronounced pattern of spots, an orange belly, no light band at the flanks and no crest on the back.

In *T. taeniatus* the colour of the male is olive-green to brown with marked black spots. In the breeding season the male shows a high wavy dorsal ridge, extending from the head to the likewise wavy tail-fin. The tail shows an orange-red and a bluish stripe.

The female has only a low ridge and a more uniform yellow-brown colour at the back and a light-yellow belly. In *T. helveticus* the male has a low crest from head to tail tip and a cutaneous fold along each side of the body, and the tail is prolonged into a thin black filament. The toes of the hind legs are webbed.

Spawning

Spawning takes place in the laboratory without any particular stimulation. The average number of eggs produced per day is about 10 in *T. alpestris* and 3 to 4 in *T. helveticus*.

Rearing Eggs and Larvae

The best temperature is 15 to 25°C (57 to 77°F). Below 10°C (50°F) the development of *T. alpestris* becomes abnormal. *T. taeniatus* hatches at a temperature of about 20°C (68°F) after 7 days; feeding starts some days later and metamorphosis takes place at 65 to 115 days, depending on food and temperature.

The rate of development in *T. helveticus* is nearly the same, that in *T. alpestris* is slightly slower. Larvae can be reared according to the directions given for the axolotl until the approach of metamorphosis. At that time the larvae should be replaced in shallow water with water plants and some islands, e.g. stones, which easily can be reached.

These stones must be cleaned regularly. During metamorphosis the animals do not feed. A difficult period comes after metamorphosis, when the larvae can often be accustomed to feeding themselves only after they have first been fed individually by hand for several weeks.

It is advisable to raise the water level gradually, so that finally the animals can rest at the surface only when sitting on floating water plants. When food is given they should be driven to the bottom.

Induced breeding, Artificial fertilization

When accustomed to an aquatic life the animals can be kept in the laboratory for years. Under suitable conditions animals reared from eggs in the laboratory can spontaneously lay fertilized eggs after one year.

With the exception of *T. helveticus* spontaneous spawning seldom occurrs in older animals. Spawning can be induced by injecting the female on two successive days 100 I.E. and the male on the second day 50 I.E. gonadotropic hormone.

The injection should be made with a short needle under the skin and muscles of the belly, pricking in anterior direction (data from Z. Stefanova, personal communication).

Transport

During the breeding period the animals should be transported in water according to the instructions given for the axolotl. At other periods the instructions given for the Spanish newt can be followed.

AMERICAN *TRITURUS* SPECIES

7'. viridesceus (- Diemictylus viridescens, common spotted newt) and *T. torosus (Taricha torosus,* California newt) are amenable to permanent laboratory conditions. Data on breeding seasons, localities etc. are given by Gadow (1901) and Bishop (1943).

After metamorphosis *T. viridescens* is strictly terrestrial, but transforms into the aquatic form after two to four seasons, when breeding begins. Fertilized eggs can be obtained in the laboratory from animals collected during the breeding season.

Outside the breeding season fertilized eggs can be obtained by injecting mature females with hypophyseal extract. See Hamburger (1960) and Rugh (1962) for details on development.

T. torosus can easily be kept in the laboratory in permanent aquatic conditions, according to the directions given for the axolotl. Sexual maturity is reached in about three years, but oviposition in the laboratory has not yet been possible.

FIRE SALAMANDER

(Salamandra salamandra)

This species belongs to the family Salamandridae and occurs in southern, central and western Europe, north-west Africa and south-east Asia. It is a terrestrial animal, which needs an airy damp environment and a temperature of about 16°C (64°F).

Tanks should be provided with damp sand, moss, niches under stones or broken flowerpots, and a shallow bowl of water into which the female can go to drop her young. The tank should be well closed with iron or plastic gauze to prevent escape.

Adult salamanders feed on earthworms and slugs, but they also take pieces of lean meat provided it is moved about gently.

The young are born alive. In nature, mating and the birth of the young takes place throughout the year except during winter. The sperm remains active in the spermatheca for a long time.

No statements could be found in the literature about the possibility of obtaining offspring from animals kept permanently in captivity. The larvae live initially in water and can be reared according to the directions given for the axolotl. They start feeding immediately after birth.

At the beginning of metamorphosis the containers should be provided with islands of moss, stones etc. During metamorphosis no food is taken, but afterwards the animals feed readily on small worms. The young should be separated from the adults to prevent them from being eaten.

Several other *Salamandra* species, with some information about rearing them in captivity, are described by Klingelhoffer (1956).

7

REPTILES

Members of the class Reptilia can be distinguished fairly easily from most other vertebrates by the fact that their bodies are covered by dry horny scales; in tortoises and their relatives these are modified to form the horny plates of the shell.

They breathe air by lungs at all stages in life and do not pass through a gill-breathing tadpole stage like many amphibians. They are said to be cold-blooded (poikilothermic), in that their body temperature usually varies with that of their immediate surroundings, whereas in the warm-blooded (homoiothermic) birds and mammals the body temperature is kept at a constant level even though that of the environment may fluctuate.

The eggs of reptiles are fertilized internally and, except in viviparous forms, they are laid on land, even though the adults (for instance, turtles) may be aquatic. In the evolutionary sense reptiles are broadly intermediate between the amphibians on one hand and the birds and mammals on the other.

The modern types of reptiles, especially crocodiles, are more closely related to birds than to mammals. They are divided into four orders: the Chelonia (tortoises etc.), Crocodilia, Rhynchocephalia (containing only the lizard-like tuatara, *Sphenodon,* of New Zealand) and Squamata (lizards and snakes).

Further information on reptiles is given in the books by Pope (1956), Bellairs (1957), Schmidt and Inger (1957), Mertens (1960), Hellmich (1962), and Vogel (1964).

The life of reptiles is greatly influenced by environmental conditions of which temperature is probably the most important. They require fairly high temperatures to keep them active, so that species which live in temperate countries must hibernate during the winter.

At the same time, they are rapidly killed by overheating and an increase in body temperatures to much over 45°C (113°F) is lethal even to most desert species; certain reptiles may die at body temperatures even below 40°C (104°F).

Many reptiles have specific preferences for particular ranges of temperature and seek out environmental conditions suitable for keeping their bodies within the optimal thermal limits.

They bask in the sun or lie on warm surfaces in order to raise their temperature to the level requisite for activity, and when in danger of overheating they shelter under vegetation, below ground or in water. In this way they are able to practise a measure of temperature-control, relying primarily on external sources of heat (such as solar radiation) for temperature-maintenance.

This method contrasts with that employed by the birds and mammals which can control and utilize the heat generated within their own bodies.

In captivity reptiles may often be denied facilities such as basking sites or shade necessary for keeping their bodies at the optimum degree of warmth. Other conditions which probably play important roles in their lives, such as humidity, light and variety of diet, may also be unsuitable.

Cyclical activities such as mating, skin-shedding and hibernation are likely to be upset in captivity. Furthermore, many reptiles in the wild show territorial behaviour during the breeding season, and this may be inhibited by lack of space and, overcrowding.

It is hardly surprising that the behaviour of reptiles in captivity is often a travesty of that in nature. They are liable to become sluggish and refuse to feed, eventually dying of starvation. They seldom breed, so that stock can only be replenished by purchase or collection of fresh specimens.

These are some of the reasons why reptiles are used less often in experimental work than members of the other vertebrate classes. Nevertheless, many species of reptiles can be kept in the laboratory without great difficulty, although the ordinary animal house, designed primarily for small mammals, may not be suitable for their accommodation.

They require special cages, a measure of individual attention and, often, living food; species which require living vertebrate food may be avoided on humane grounds. Some familiarity with the habits of the particular species used is always desirable.

USES OF REPTILES IN THE LABORATORY

Most live reptiles which enter laboratories in England are probably killed as material for students to dissect, but attention should be given to problems of packaging and transport not only on grounds of humanity but also because wastage from death may otherwise be high.

Reptiles are occasionally used in research in the same way as small mammals and some amphibians-purely as convenient creatures in which to investigate some biological process. For example, they are suitable for certain types of experimental work on the nervous system or, in the case of lizards, for studies on tail regeneration.

In such work the small size of many species is an asset, since it facilitates the preparation of serial microscopic sections through quite large regions of the body.

Because of their remote relationship to man and domestic animals, reptiles are, however, little used in medical and veterinary research except, of course, where the study of snake venoms is concerned.

The majority of biologists who work on reptiles have some primary interest in this group of animals as such, and are willing to take considerable trouble over their care and maintenance.

For many types of experimental research it is only necessary to keep reptiles in the laboratory for a matter of a few weeks or months in a state of health. With many species this is not a very difficult problem so long as they can be induced to feed.

No great attempt to reproduce a semblance of natural conditions is necessary and indeed it is probably better not to make one. Neither is it advisable to allow the animals to hibernate; they can be kept active and feeding at a more or less constant temperature throughout the year.

Individual reptiles may survive for many years in captivity. The directions given in this chapter, however, are mainly intended for those who wish to keep them for limited periods for scientific work, rather than for vivarium-keepers or zoo curators.

THE DIFFERENT KINDS OF REPTILES

The kinds of reptiles most often used in English laboratories are probably the European tortoise *(Testudo graeca)*, the common and green lizards *(Lacerta vivipara* and *L. viridis)* and the grass snake *(Natrix natrix)*. The common lizard, the slow-worm *(Anguis fragilis)*, the grass snake and the adder *(Vipera berus)* are inhabitants of Britain and can be obtained, during the summer at least, from local collectors.

The European tortoise and various water tortoises or terrapins from Europe and North America, and also the green lizard, a common continental form, can be purchased in season from many pet-shops, but may be very difficult to obtain in the winter.

A wide range of other species, including crocodilians and pythons, can be obtained from specialized animal dealers. Useful notes on the more readily available types are given by Leutscher (1961).

ORDER CROCODILIA

This contains about 25 species of crocodiles, alligators, caimans and gharials. The different types vary only slightly in habits and appearance as, for example, in the relative length and width of the snout.

They reproduce by laying eggs, in either craters dug in the sand or nests built of vegetable debris. The species most often sold are the broad-fronted caiman *(Caiman latirostris)* from South America, the American alligator *(Alligator mississippiensis)* and the Nile crocodile *(Crocodylus niloticus)*.

The animals are usually juveniles of under 3 ft (0·9 m) long. As they may all be sold as 'baby alligators' it is advisable to check their identity. The Nilotic and most other kinds of crocodile have relatively long snouts, and the teeth in both jaws are visible when the mouth is closed. The large fourth tooth on each side of the lower jaw fits into a notch in the upper, and is conspicuous.

In alligators and caimans, on the other hand, the lower teeth are more or less hidden when the mouth is closed; the fourth lower teeth fit into pits in the upper jaw and are invisible. The broad-fronted caiman has an even shorter, wider snout than the alligator and its tip is upturned.

As in certain other caimans, its eye-sockets are connected in front by a curved bony ridge which is absent in alligators. For further details, a key such as that by Wermuth and Mertens (1961) should be consulted.

Because it is comparatively unaggressive and resistant to low temperatures, the American alligator is the most suitable crocodilian for laboratory work.

ORDER CHELONIA OR TESTUDINATA (TORTOISES etc.)

This order contains about 250 living species and nine or more families, depending on the system of classification used. They are found throughout the warmer parts of the world. Broadly speaking, chelonians can be divided into three groups on the basis of habits, although this arrangement does not conform with strict zoological classification.

Those which are terrestrial are usually called tortoises; they have high domed shells and strong claws and are almost exclusively herbivorous. The famous giant tortoises of the Galapagos and Indian Ocean islands belong to this group.

Amphibious, freshwater chelonians are often called turtles or terrapins; they usually have rather flat shells and webbed or partly webbed feet, and are more or less carnivorous.

Finally, there are the almost entirely aquatic marine forms known as marine turtles, such as the green turtle *(Chelonia mydas)* and the hawksbill *(Eretmochelys imbricata)*, the source of commercial tortoiseshell; these have paddle-like limbs and feed on both animal and vegetable material.

These sea turtles are unlikely to be used in the laboratory and must be accommodated in sea-water aquaria. The names 'tortoise', 'terrapin' and 'turtle' are not used consistently, however, and American writers may apply the name 'turtle' to any type of chelonian.

In the United States the term 'terrapin' is generally restricted to certain of the smaller species which are sold as food, such as the diamond-back terrapin (genus *Malaclemys)*.

The structure of chelonians is greatly modified in connection with the shell and they are in many ways the most specialized of modern reptiles. They have no teeth, but a horny beak instead. They reproduce by egg-laying, the eggs of tortoises being hard-shelled while those of some terrapins and turtles have a leathery or parchment-like shell.

The eggs are usually laid in holes dug in earth or sand by the female, the nest of some species being moistened with urine. It is often possible to distinguish the sex of a chelonian by external features. In males the tail is generally long and thick, while in the female it is relatively small.

In male tortoises the plastron or under-part of the shell tends to be concave in shape, fitting over the carapace or upper shell of the female when the male mounts the female in coitus. In some species there are also marked sexual differences in size and other characters. In chelonians, as in crocodilians, the male has a single penis.

Many species show some form of courtship behaviour. For example, a male tortoise will follow the female about during the breeding season, butting her shell and biting at her feet. Some terrapins of the genera *Chrysemys* and *Pseudemys* have an elaborate courtship display, the male swimming backwards in front of the female and stroking her face with his elongated front claws.

Chelonians are traditionally long-lived and even some of the smaller species have been known to survive for periods of over 50 years.

The land tortoise most often imported to England is the common European species, *Testudo graeca,* a native of southern Europe and North Africa. It belongs to the family Testudinidae. Large individuals have a carapace nearly a foot long.

This animal is sometimes called the 'spur-thighed tortoise' because it possesses a prominent spur on the back side of each thigh; the related Hermann's tortoise *(T. hermanni),* with which it is easily confused, lacks these spurs, but has a horny tubercle on the tip of the tail instead. The box-turtles *(Terrapene)* of North America are another group of small chelonians which are in some ways intermediate between the typical land and freshwater forms.

They owe their name to the fact that they have a flexible hinge across the plastron and, by retracting their extremities and elevating the two plastral segments, they can box themselves up very completely in their shells.

The terrapins most commonly available are the European pond-terrapin *(Emys orbicularis)* from central and southern Europe and North Africa, and the North American redeared or elegant terrapin *(Pseudemys scripta elegans);* babies of the latter are often sold but need special care.

Both species belong to the family Emydidae. The European form has a blackish carapace, sometimes 8 in long; the shell, head and limbs have small yellow streaks or dots. The red-eared terrapin is more handsomely marked, having a prominent red or orange stripe on the side of the head behind the eye and yellow markings on the shell and extremities.

This terrapin seems very suitable for laboratory work, and directions for keeping it are given by Boycott and Robins (1961). The species *Pseudemys scripta* contains various other sub-species, such as the yellow-bellied terrapin (P.s. *scripta)* which may also be on sale. The painted terrapin *(Chrysemys picta),* another related form, has bright red markings along the edge of the shell.

Larger forms which might be kept in the laboratory are the snapper *(Chelydra serpentina;* family Chelydridae), a fierce predatory terrapin from North America, and some of the soft-shelled turtles (family Trionychidae); these are adapted for life at the bottom and have flat shells and long necks which can be darted out at great speed to capture prey.

They have considerable powers of underwater respiration, small vascular papillae in the throat serving as a kind of gill. The handbooks by Carr (1952) and Wermuth and Mertens (1961) are helpful in the identification of chelonians, which may present some difficulty, especially in the case of the small North American terrapins. The little book by I. and A. Noel-Hume (1954) is a useful and not too technical guide to identification, and also contains much information on keeping these and other species.

ORDER SQUAMATA (LIZARDS AND SNAKES)

Lizards are classified in the sub-order Lacertilia or Sauria, snakes in the sub-order Ophidia or Serpentes. The two groups are closely related and the fact that many lizards, for example the slow-worm, are limbless sometimes makes distinction difficult.

It may be stated as a rough guide that a limbless reptile with a single row of large scales down the belly in front of the vent and a transparent spectacle-like eye-covering is a snake, whereas one with small belly scales and movable eyelids is a lizard.

There are some cases, however, in which this distinction breaks down. The lizards and snakes are by far the largest groups of modern reptiles, containing each some 2,500 species or more, and about 20 and 10 families respectively.

They are world-wide in distribution; the great majority of species are of course found in warm countries, but two, the common lizard *(Lacerta vivipara)* and the adder *(Vipera berus),* range as far north as the Arctic Circle. Lizards and snakes are distinguished from other reptiles in having the anal opening or vent set in the transverse instead of the longitudinal plane, and in the possession by the males, of paired organs of copulation known as hemipenes.

In all snakes and many lizards the tongue is protrusible and its tip is forked to a greater or lesser extent. The tongue acts in conjunction with a pair of important sense organs, the organs of Jacobson, which lie above the roof of the mouth.

The sense which they serve is akin to smell, scent particles being carried to them by the tongue. The mechanism explains the constant tongue movements made by snakes and certain lizards when they are exploring their surroundings or investigating their food. Most lizards and snakes lay eggs with parchment-like shells.

These are usually deposited in holes or under vegetation and are hatched by the warmth of their surroundings, but in a few species, for example some pythons, the eggs are brooded by the mother. A substantial number of both snakes and lizards produce their young alive, retaining the eggs within their bodies until development of the embryos is complete.

Mating behaviour shows much variation; certain lizards, notably those in the agamid and iguanid groups, go through elaborate forms of courtship display in which the erection of crests and other appendages, bobbing movements of the head and fore-quarters, and colour change may play a conspicuous part.

Sex distinction is often possible on the basis of size or colour, especially in lizards, but may be difficult in snakes. In males the base of the tail may appear somewhat swollen owing to the presence of the hemipenes.

Most lizards shed their skins periodically in large flakes, but in snakes the skin is usually shed as a single piece or slough. The snake begins by rubbing it off at the nose against some firm object and then creeps out of it, often turning it inside out.

The frequency of moulting varies with the age and health of the animal, but may be several times a year. The skin of a snake goes dull a week or more before it is shed, and the transparent spectacle over the eye becomes bluish and opaque, regaining its transparency a few days before the act of moulting.

Snakes usually stop feeding during the pre-moulting period, when their sight is impaired. At such times normally tame animals may become irritable and liable to bite. Captive snakes which have difficulty in shedding their skins should be helped by hand, and care should be taken to see that the old spectacle is detached.

Many of the smaller lizards and snakes will live for 5 to 10 years or more in captivity, and some of the larger kinds such as pythons have been known to survive for over 20 years.

The different kinds of these reptiles vary enormously in appearance and habits. Here it is only possible to mention some of the more important types and to indicate those species which seem most suitable as laboratory animals.

Lizards

Family Geckonidae (Geckos)

Common genera: *Hemidactylus, Tarentola, Gecko, Phyllodactylus,*

Sphaerodactylus, Stenodactylus. This is a large family of small lizards, very abundant in warm countries throughout the world. Many geckos are adapted for climbing and possess specialized digits which enable them to cling to smooth surfaces, even upside down.

They are often nocturnal or crepuscular, but some species bask in the sun. A few geckos are terrestrial, some living in deserts. The great majority reproduce by egg-laying, the eggs being unusual in having hard shells. The tail is generally very fragile and regenerates readily. Most geckos thrive in captivity and do not require direct sunlight.

Family Agamidae

Common genera: *Agama, Uromastix, Calotes, Amphibolurus (e.g.* bearded lizard), *Chlamydosaurus* (frilled lizard), *Draco* (flying lizard). This is a large group of Old World lizards. Many types are desert-living, some are arboreal and a few amphibious; they are mostly diurnal and sun-loving.

Some agamids have special pouches, frills etc., around the throat which are erected by the males during courtship and threat display. The great majority are egg-layers. The more active species are not easy to keep in the laboratory.

Family Iguanidae

Common genera: *Iguana, Cyclura, Basiliscus, Anolis, Phrynosoma, Sceloporus, Crotaphytus, Uta, Sauromalus.* This important family is restricted to the New World and Madagascar. There are some large (2 to 5 ft long) partly arboreal, partly amphibious genera such as *Iguana* and *Basiliscus* and many smaller terrestrial, often desert-living ones such as *Phrynosoma* (horned lizards), *Sceloporus* (swifts) and *Sauromalus.*

The *Anolis* lizards are small tree-dwelling forms with a distensible dewlap or throat-fan and have striking powers of colour-change. They are often called 'chamaeleons' in the U.S.A. and have often been used in laboratory studies. Many iguanids show a curious parallel with agamids in appearance and habits, and resemble them in showing marked territorial behaviour and courtship display. The majority are egg-layers.

Family Chamaeleonidae

The chamaeleons are specialized arboreal lizards with highly developed eyesight, and powers of colour-change, prehensile limbs and tail, and long, extensible tongues with which they capture their insect prey. They are restricted to the Old World, one species *(Chamaeleo chamaeleon)* being found in southern Europe.

The majority are egg-layers. Chamaeleons do better in captivity than is often believed if given proper attention and sufficient food, and may even breed. Some of the dwarf species such as *Microsaura pumila* from southern Africa, a viviparous type, might well prove fairly easy to maintain in the laboratory.

Family Lacertidae

This consists of small or medium-sized, predominantly terrestrial lizards and includes the common lizard *(Lacerta vivipara),* the sand lizard *(L. agilis),* also found in England, the wall lizard *(L. muralis),* the green lizard *(L. viridis),* and the eyed lizard *(L. lepida)* which may reach a length of 2 ft.

The last-named three species are European; the green lizard is the easiest to keep in the laboratory and its size (10 to 18 in) makes it suitable for many types of experimental work. The common lizard is less easy to keep indoors than might be expected, and seems very prone to mite infections which are often fatal.

Lacerta dugesii, the Madeira wall lizard, may also be a useful laboratory species and will usually feed throughout the year. Most lacertids reproduce by egg-laying, *Lacerta vivipara* being an exception. The eggs of this form will develop in culture after being removed from the mother, and can be subjected to some experimental procedures.

Family Scincidae

Common genera: *Scincus, Chalcides, Mabuya, Tiliqua, Egernia.* This is a widespread family of small and medium-sized lizards, some forms *(Chalcides)* being found in Europe. Skinks characteristically have long bodies and short legs; in many species the limbs are reduced or absent. The scales are often smooth and shiny.

Most skinks are secretive or burrowing in habits and a number live in sand. Some Australian forms such as the bluetongue *(Tiliqua scincoides)* and the stump-tail or shingle-back *(T. rugosa)* are large stout lizards 16 in (40 cm) long. These and many other skinks live well in captivity. A large number of skinks are viviparous and some have well-developed placentae.

Family Anguidae

This family includes limbless forms such as *Anguis fragilis* (the slow-worm) and *Ophisaurus* (American and European glass-snakes), and also some such as *Gerrhonotus* (alligator lizards) in which limbs are present. They are all terrestrial, the limbless species showing burrowing tendencies. The slow-worm is very easy to keep and has been known to breed in captivity.

It requires little sun or space and has simple though definite food preferences. Survival for over 50 years is recorded, though this is probably exceptional. It is viviparous and its eggs can be cultured after removal from the mother.

Family Varanidae

The monitor lizards are a large family of lizards found in the subtropics and tropics of the Old World. Most of them are large, *V. komodoensis* (the Komodo dragon) being the largest living lizard with a maximum length of about 10 ft (3 m). Some species, such as *V. salvator* and *V. niloticus,* are amphibious but also spend some time in trees.

They are highly predacious, feeding on eggs, carrion, and any animals which they can kill. Their size makes them suitable for some kinds of laboratory work and their skeletons are useful for class demonstration. Although often vicious they usually feed well in captivity once they have settled down, and may live for many years. They are all oviparous.

Family Helodermatidae

This contains the only two poisonous lizards known, the gila monster *(Heloderma suspectum),* from parts of the southern U.S.A., and the larger Mexican beaded lizard *H. horridum.* Anyone intending to work on these lizards should consult the fine monograph by Bogert and del Campo (1956).

Snakes

The following groups are of most interest here

Family Boidae (Boas and Pythons)

These snakes retain vestiges of the hind limbs which are visible in some species as small claws on either side of the vent. All are powerful constrictors, the prey being seized in the long recurved teeth and quickly enveloped in the coils.

Death is generally caused by suffocation or stopping of the heart. These snakes are not poisonous and some will become tame in captivity. Snakes of this family vary in size from burrowers 2 to 3 ft long, such as *Eryx* to the very large anaconda *(Eunectes murinus)* and reticulated python *(Python reticulatus)* which may reach 30 ft (9 m).

Pythons lay eggs which are brooded in the coils of the female, while most boas give birth to living young. Pythons are found in the Old World and Australia, boas mainly in the New World.

Family Colubridae

Some common genera are listed below in Table elsewhere in this chapter. This family contains the great majority of snakes, including most of the familiar harmless types.

It is often divided into two groups, the aglyphs, which have simple teeth, and the opisthoglyphs, in which one or more of the posterior maxillary teeth are grooved to transmit venom.

Aglyphs, such as the grass snake, are for practical purposes non-venomous, their prey being killed by constriction or eaten alive. The opisthoglyphs or back-fanged snakes are usually described as mildly venomous, but at least one species, the boomslang from southern Africa, has been known to kill human beings. A bite from one of the larger opisthoglyphs should be treated as serious.

Family Elapidae

Common genera : *Naja* (cobra), *Bungarus* (krait), *Dendroaspis* (mamba). In these snakes the front teeth of the upper jaw are modified to form fangs and possess canals down which the venom flows. The majority of this group are highly poisonous. All the Australian venomous snakes are elapids.

Family Hydrophidae (sea-snakes)

These are highly poisonous snakes, found mainly in the warm oriental seas. Most species are entirely aquatic and give birth to living young in the water. They are kept for purposes of venom-extraction in a research centre at Penang and will live for a while in tanks of sea-water. Notes on keeping them are given by Reid (1956).

Family Viperidae

Common genera: *Vipera (e.g.* adder), *Bitis (e.g.* puff adder), *Crotalus* (rattlesnakes), *Agkistrodon* (moccasin, copperhead). This group is divided into two sub-families, the Viperinae and the Crotalinae or pit-vipers such as the rattlesnakes; the pit-vipers are so called because they possess a pit on either side of the head between the eye and the nostril which functions as a heat receptor.

In all Viperidae the maxillary bones are hinged to the skull and can be rotated in such a way as to erect the fangs which are fixed to them. Each fang is traversed by a canal through which the poison from the venom gland flows into the wound.

The fangs of these snakes may be very long and when not in use they are folded back along the roof of the mouth. It shoulbe d emphasized that there is no easy way of distinguishing harmless snakes in general from poisonous ones without examining the teeth, and all unidentified snakes should be treated with caution.

Since the poison fangs of snakes, like reptilian teeth in general, are renewed throughout life, fang-extraction is only a temporary method of rendering a snake harmless. A more permament method is to block the venom duct by coagulation with an electric cautery; this technique has been used successfully on rattlesnakes and apparently has little effect on their general health.

Generally speaking, however, the use of live poisonous snakes in the laboratory is inadvisable unless their presence is really necessary, as in venom research.

HUSBANDRY

Handling

Most reptiles will bite if they are unaccustomed to handling, but with the smaller non-poisonous species such injuries seldom have untoward effects. Large lizards such as monitors, the bigger terrapins and snakes such as pythons can all produce severe wounds which require treatment; in the case of snakes the recurved teeth may break off in the wound and must be removed.

The secret of handling reptiles is to grasp them firmly, confidently and rapidly; a hesitant approach often provokes a bite. Potentially vicious reptiles such as monitors, small crocodiles and the bigger non-poisonous snakes should be held firmly behind the head; it is important that the weight of the creature's body should be supported comfortably by the other hand or, in the case of large reptiles, by assistants who can prevent the body and tail from thrashing.

The jaws of crocodilians may be tied or taped together if necessary; their jaw-opening muscles are comparatively weak. Leather gloves afford a useful protection from bites and scratches. Lizards prone to shed their tails should not be seized by these members.

Poisonous snakes, even when apparently tame, should be handled as little as possible and with great circumspection. If possible they should be lifted by means of a snake stick. This is a rod made of bamboo or similar strong but light material with the tip padded with felt.

A soft leather strap passes over the padded end through a metal sheath attached to the handle to form a loop. This is passed over the snake's head and drawn tight by pulling on the handle. Great care must be taken when this method is used, as the snake may easily dislocate its neck by struggling.

Figure 7.1: Snake rod for lifting and pinning down heavy snakes.

To prevent this, its body should be held and supported as quickly as possible after it has been noosed. For heavy, sluggish, viperid snakes, such as puff adders and rattlesnakes, the use of this type of snake stick is not advisable.

These snakes are best lifted by means of a strong rod with a metal support at right-angles at the end. This can usually be slipped under the snake's body about half-way down so that it balances when lifted from the ground.

When these large vipers have to be handled, the head can be firmly held down with the same stick and the snake gripped behind the head with the free hand, Figures elsewhere in this chapter. It must be remembered that these snakes have extremely long fangs and the fingers gripping the snake should be kept well clear of the mouth.

The snakes are very strong and if loosely held may twist round and use one fang by biting sideways. Some of the smaller vipers, such as the European adder, *Vipera berus,* may be lifted by the tip of the tail and carried for short distances at arm's length.

If not watched, however, they may be able to obtain a purchase on their own bodies and climb up to bite. It may be convenient to immobilize a snake for purposes of examination or operation by inducing it to enter a transparent tube. Such a tube can be made by rolling a sheet of cellulose acetate and securing it with tape or elastic.

Holes can be cut in the walls of the tube to allow access to the snake's body. If one corner of a sack is cut away and the edges of the opening are fastened round the mouth of the tube, the snake can be dropped into the sack and will probably enter the tube of its own accord. A similar but more elaborate procedure is described by Pope.

Identification of Individuals

It is often necessary to identify individuals among a number of animals kept in a single cage, and this cannot always be done on the basis of such features as size and colour. Marking with spots of paint or quick-drying enamel is useful as a short-term means of recognition, but such spots may be rubbed off, or shed when the animal moults.

Chelonians can easily be marked by filing a notch in the edge of the carapace; this will remain visible for many years. Lizards and crocodilians can be identified by clipping off the outer phalanges from one or more of the toes with sharp scissors. Careful notes should be kept of the position of the extremities clipped (e.g. right hind foot, 1st digit).

This is a more or less permanent method often used in field studies. If any regeneration does occur it is slow and often incomplete. Snakes can be marked by clipping the ventral scales. Other methods of identification, such as tattooing, have been recommended.

First Aid in Cases of Snake-bite

Excess venom should be wiped off the skin which should be thoroughly washed. The wound should be sucked, either with a suction cup or by the mouth; the latter should do no harm provided there are no abrasions to the mouth or lips. In the case of bites by very dangerous vipers and rattlesnakes the wound should be incised to its full depth to promote bleeding and suction should then be applied.

This measure is of more doubtful value in the case of bites by elapid snakes whose venom is absorbed very quickly. In the case of snakes which are not very poisonous, such as the European adder, incision may do more damage than the bite and is not recommended.

A tourniquet should be applied 2 to 4 in above the site of a bite on a limb, between it and the heart, in order to localize the effects of venom. It should be tight enough to block the superficial venous and lymphatic return, but not to block the arterial flow. It should be released for about 11 minutes every 10 minutes.

Exercise should be avoided and the patient put to bed. A bitten limb should be immobilized by splinting. Treatment for shock or respiratory paralysis may be necessary, and medical attention should be obtained. The application of potassium permanganate to the wound, and the drinking of alcohol, are not advisable.

It is most important that in the case of bites by seriously poisonous snakes antiserum should be given as soon as possible and a stock of this, in an active state, should always be available in a laboratory where poisonous snakes are kept.

Antiserum should be given by intramuscular or intravenous injection. In view of the occasional dangers of serum therapy, bites by snakes such as the adder, which are not very poisonous, may be best treated by conservative measures such as rest in bed and immobilization of the bitten part.

BREEDING

Although reptiles are unlikely to complete a full breeding cycle in the laboratory, animals inseminated before capture may lay eggs or produce their young.

In certain forms, notably some turtles and snakes, sperm can be stored alive for long periods within the reproductive tract of the female and there are instances of fertile eggs being laid after months and even years of isolation.

Many ways of artificially incubating reptile eggs have been suggested. They should be kept neither too dry nor too moist. In the latter case they are prone to fungus infection; in the former they become desiccated and collapsed.

The eggs should be shallowly imbedded in a medium of clean, slightly damp sand or sawdust and placed in a clean container. Zweifel (1961) who reviews the various methods, recommends keeping them in damp sand in a polythene bag, slightly inflated and sealed with a rubber band. Temperatures of 25 to 30°C (77 to 86°F) seem suitable for the eggs of most subtropical and tropical species.

Many reptiles have an incubation period of two to three months, the duration being prolonged at lower temperatures. Once the eggs have been placed in their artificial nest they should be disturbed as little as possible; except in the case of reptiles such as pythons which brood their eggs, the eggs should always be removed from the parent's cage.

ACCOMMODATION

A vivarium set up with rocks and plants looks attractive and is worth setting up if reptiles are to be kept as pets or for exhibition. An attempt to simulate the animals' normal surroundings may be necessary if some aspect of its natural behaviour, such as courtship, is to be studied or in attempts at breeding.

Elaborate cages, however, require much time and trouble to maintain and if not kept really clean may harbour parasites. Reptiles kept for most laboratory purposes should, therefore, have cages or enclosures which are as simple as possible.

Outdoor Enclosures

If there is space in the laboratory grounds, British and some temperate-zone reptiles can be kept out of doors, at least in the summer. The additional space available and the possibility of the animals obtaining some natural food such as insects are advantages, and breeding is more likely to occur.

On the other hand, snakes and lizards kept in this way tend to become very wild, darting for cover when approached, and in winter problems of hibernation arise. British species can be left to hibernate in deep well-drained holes filled with loose earth and covered over, if not required during the winter.

Continental reptiles should probably be brought indoors and either allowed to hibernate in a cold room or shed under frost-proof conditions or in boxes filled with straw, or kept active artificially at raised temperatures.

The continental reptiles most suitable for keeping out of doors are the European tortoise *(Testudo graeca)* and the green and wall lizards *(Lacerta viridis* and *L. muralis)*. Both of the latter have been artificially introduced into this country at various times, with varying measures of success.

A wire-netting fence about 1 ft (30 cm) high makes a suitable pen for tortoises, and can be erected on a lawn. It should contain a dish of water deep enough to allow the tortoise to immerse its head when drinking and sunk to the level of the ground.

There must also be some kind of waterproof shelter into which the tortoise can retreat from wet, cold, or excessive heat. This can be made from a wooden box roofed with linoleum.

Outdoor cages for lizards can be made out of four sheets of glass or any really smooth material, the ends of each sheet being let into grooves in wooden uprights. Triangular pieces of glass should be placed over the tops of each corner and fixed on the inside with Sellotape.

The enclosure as a whole may be placed on a smooth surface such as a flat roof or strip of concrete, or it may be dug a couple of inches into the ground. Good drainage is essential so that the floor of the pen cannot become waterlogged.

The height of the glass should be at least twice the length of the lizard (including tail) and no vegetation which might help the animal to climb out should be allowed to grow near the sides. Tins with perforated bottoms and sunk in the ground are useful repositories for meal-worms and other living food.

A shallow dish of water, sunk to ground level, should always be provided, but this must be easily climbable by the lizard. There should also be some kind of retreat such as a dry wooden box filled with straw which is easily moved and opened.

Another, less sightly, type of enclosure can be made by bolting together sheets of galvanized metal to form a circle. Permanent enclosures made of glazed bricks or smooth concrete with an overhanging lip of tiles are ideal, but any form of reptiliary with opaque walls must be large enough to allow some part of it to receive sun during the greater part of the day.

Because of their great length, and consequently greater climbing ability, snakes are less easy to keep in outdoor enclosures than small lizards. A smooth vertical surface somewhat higher than the length of the snake will probably prevent escape, but special precautions must be taken with poisonous species.

All outdoor vivaria should be protected by netting or some other material from cats, birds and other predators, which may do great harm to a collection. Reptiles in hibernation are liable to attacks by rats.

It may finally be emphasized that the essential requirements for all outdoor enclosures for reptiles are good drainage and a sufficiency of suitable basking sites on one hand and shelter from cold, wet, and powerful sun on the other. Generally speaking, snakes require less sun than lizards (except the slow-worm and other burrowers) and tortoises, but more water.

Indoor Cages

Terrapins and crocodilians should be kept in covered aquarium tanks partly filled with water; a depth of 2 to 3 in (5 to 7·5 cm) is adequate for small specimens. About half of the space inside the tank should be converted into land.

Bricks placed on each side of the tank to support a board just above water-level are very suitable since the animals often like to lie under cover in the water. It is not necessary to put sand or shingle in the tank, except for bottom-living terrapins such as *Trionyx,* which bury in it.

European and North American terrapins and alligators should be kept at temperatures of between about 24 and 30°C (75 and 86°F), crocodiles and caimans between 27 and 30°C (81 and 86°F). The temperature may be slightly reduced at night.

Most of these reptiles like to bask at times, and the cage should if possible be reached by some direct sunlight. If this is impossible, however, the deficiency can be remedied by fitting one or more electric-light bulbs into the top of the tank; the reptiles should not be able to make contact with them, or they may burn themselves.

If the tank is suitably enclosed, the presence of the electric bulbs may be sufficient to maintain the temperature; otherwise a thermostatic heater should be used.

Terrapins, in particular, are messy feeders, generally eating beneath the water. The tanks must therefore be cleansed constantly and this is a great deal easier if they are fitted with plugs. It may save trouble to transfer the animals to a separate tank when food is given.

Small lizards and snakes can be kept in containers of almost any kind, provided they are light and have some ventilation. Access to direct sunlight for part of the day is desirable, though perhaps not essential, for many lizards, but under no circumstances should cages be placed where there is likely to be danger of overheating, as in a glass window facing south.

Electric bulbs can be used to simulate sunlight and maintain the temperature; ultraviolet light may perhaps be beneficial but is probably unnecessary for animals kept for most laboratory purposes.

Snakes require less light than lizards, apart from nocturnal species such as many geckos, and burrowing or secretive forms such as slow-worms. Temperatures of 25 to 32°C are generally suitable for most subtropical or tropical forms.

Wood is probably the best material for the cages, sheets of glass being let in at the front and sides. Parts of the top and back at least should be covered only with perforated zinc, and a small hole closed by a cork may be bored in the top to allow food to be dropped in without risk of lizards escaping.

Furniture inside the cage should be minimal. Cover may be given by pieces of wood or stone, hollow bricks or corrugated asbestos. Climbing reptiles such as chamaeleons and tree-snakes require a few twigs or branches; geckos may be given a few vertically placed pieces of bark.

A dish of water should be provided, and in the case of snakes this should be large enough to allow the animals to immerse themselves; many snakes, in addition to species which are normally amphibious, like to bathe, especially before shedding their skins.

Snake cages should also contain a brick or some other rough heavy object against which they can rub off the old skin when they are moulting. The floor should be kept clean and as dry as possible; hence the advantage of wood or some other absorbent material.

Slight tilting of the cage may be desirable to enable any water which may slop out of the dish to drain to one end. Sheets of newspaper placed on the floor are absorbent and also facilitate the business of cleaning. Sand, leaf mould, moss and other such materials are best used only in the cages of burrowing species such as skinks or slow-worms.

The latter live well in covered glass bowls or jars if provided with leaf mould, a small dish of water, and a piece of wood or bark under which to hide.

Neither snakes nor lizards require a great deal of space in the laboratory. A cage about 2 ft (60 cm) long and 1 ft wide will accommodate three or four green lizards or a couple of snakes 2 to 3 ft long, but the crowding of large numbers of individuals together for any length of time should be avoided.

Baby specimens should not be kept with adults and should be transferred to smaller glass containers from which there is no risk of escape.

Useful directions for keeping snakes under simple conditions are given by Littleford and Keller (1946) and by Smith (1953).

Nutrition

Most reptiles, especially snakes, can go without food for long periods, and although they should normally be fed at regular intervals a fast of a week or so (as during a holiday period when the laboratory is closed) will probably do them no harm.

At such times the temperature of their cages may be lowered slightly to reduce activity. Tortoises will eat lettuce, cabbage, dandelions, clover, grass, tomatoes and other fruit, and are very fond of yellow flowers.

Some tortoises will take raw fish or meat. Individuals often show preference for one particular food, and it may be necessary to experiment with a newly acquired animal in order to determine its choice.

Terrapins can be given chopped meat, including heart and liver, fish, earthworms, mealworms and other insects, as well as freshly killed small frogs and tadpoles if available.

Many species also like vegetable food such as lettuce which can be dropped into the tank. Young animals should be fed at least every other day, older specimens about twice a week. Small crocodilians will accept much the same food as terrapins but will not eat plants.

Reluctant feeders may be induced to eat by one or more of the following expedients: (a) slightly raising the temperature of their tank; (b) waving strips of food close to their mouths or (c) placing small pieces of food between their jaws.

In captivity chelonians and crocodilians, especially when young, are prone to develop deficiency diseases resembling rickets; these may lead to softening of the limb bones and, in the case of chelonians, of the shell. The diet may therefore be supplemented with fishliver oil smeared on the food or placed in the back of the mouth.

Extra calcium should also be provided for amphibious forms by placing so-called bones of cuttle-fish, pieces of plaster of Paris, or chips of vertebrate bone in their tanks.

The majority of lizards feed on any small creatures which they can overcome, and should be given as much variety of food as can be obtained.

Mealworms (larvae of *Tenebrio* beetles) and gentles (fly larvae) are most valuable as a stand-by, and can be placed in the lizards' cages in a petri dish to stop them crawling about; they seem to be inadequate, however, as the only food for long periods, and some lizards refuse to touch them after a time.

Earthworms, slugs, flies, butterflies, moths and their larvae, cockroaches, grasshoppers, spiders and centipedes should be offered when possible; grasshoppers and spiders are particularly relished by many species. Locusts and various other insects can be raised in the laboratory and advice about insects which might be worth breeding as reptile food may be sought from entomologists.

Many insects can easily be caught during the summer by shaking bushes over a tray or cloth, by drawing a net through vegetation or by the use of a moth trap.

Some of the larger lizards such as eyed lizards *(Lacerta lepida)* and monitors *(Varanus)* will eat young mice and smaller lizards. For the larger skinks, such as *Tiliqua,* and for heloderms, chopped meat and raw egg is recommended. Slow-worms will feed entirely on earthworms and slugs, especially the white *Agrolimax*.

Not many lizards eat ants, but the small spiny agamid *Moloch horridus* of Australia feeds exclusively on certain species of these insects. Baby lizards, including *Lacerta vivipara,* may prove difficult to feed, but can be offered *Drosophila,* aphids (e.g. greenfly) and tiny spiders.

A few lizards, such the agamid *Uromastix* and some of the big iguanas, are more or less herbivorous and should be given lettuce, fruit and other plant food along with insects.

Even some predominantly insectivorous species, such as green and wall lizards, will often eat pieces of tomato, grape and other fruit.

All snakes feed on animals. Most eat vertebrates; some young snakes and the adults of a few species mainly on invertebrates. Constricting and poisonous snakes usually kill their prey before eating; other species, such as the grass snake *(Natrix natrix),* in the wild state swallow it alive.

Animals are usually eaten head first, so that their limbs lie flat against their bodies and swallowing is facilitated. They are always swallowed whole, a procedure which is made possible by the looseness of the snake's jaws and the distensibility of the skin of its head and neck.

The swallowing process may be very long-drawn-out if the prey is large; while it is going on the snake's glottis may be protruded from its mouth to prevent its breathing being obstructed.

Many colubrid snakes, including those of the genus *Natrix,* which are particularly suitable as laboratory animals, feed almost exclusively on amphibians, especially frogs and newts, and fish. Individuals of some species, such as the grass snake *(N. natrix)* also take toads.

Garter snakes *(Thamnophis)* eat earthworms, and in captivity can be given strips of raw meat and fish. Contrary to popular belief, young grass snakes may not often eat invertebrates; they feed readily on the tadpoles of frogs, toads and newts.

Although in the wild snakes seldom, if ever, eat animals which have been previously killed, they will generally take dead food in captivity. Live food such as rats or mice have been known to attack and seriously damage reptiles.

Captive snakes will usually feed by day or night, but it may be advisable to offer food to newly caught nocturnal species in the evening. The food should be removed in the morning if still uneaten.

After feeding, the temperature should be maintained at a steady level and the snake should not be disturbed since excitement or sudden cooling may cause it to disgorge its food. If possible snakes should not be handled for 48 hours after they have eaten, and it is inadvisable to feed them immediately before a journey.

Food need not always be offered freshly killed; at the London Zoo wild rabbits which have been thawed out after many weeks in cold storage have been accepted by pythons and puff adders.

A reluctant feeder may be tempted if dead food is gently moved by means of a stick or wire to give the appearance of life. Warming the carcasses of birds and mammals which have been dead for some time may also act as an inducement to feed.

Two or more snakes may attack the same food, and in their efforts to eat it one of them may damage, or even swallow, its companion. Poisonous snakes often strike at one another at feeding time, and the larger vipers may inflict serious wounds with their long fangs. Although snakes have some immunity to the venom of their own species, they may quickly succumb to a bite from another species.

Healthy snakes may do without food for several months without harm. A fast may sometimes be terminated by raising the temperature, by giving the snake a warm bath, or by repeatedly presenting different types of food in various ways. Changing the snake to a new cage will sometimes start it feeding.

If it is necessary to preserve a snake which persistently refuses to eat, force-feeding may be employed. A snake will sometimes eat if its jaws are gently opened and food is pushed into the mouth. Great care should be taken not to damage the jaws.

If it still refuses to swallow, more radical measures must be adopted. The food should be pushed with a blunt instrument for some distance down the snake's throat; any tendency to disgorge may be restrained by a ligature which should be released after 12 hours.

Dipping the food in egg will help to lubricate it and assist its passage down the gullet. In force-feeding it is important to use food which is considerably smaller than that usually eaten by the snake. It is very difficult to force-feed small snakes. If other food is not available the snake may be forcibly fed on chopped meat.

The administration through a pipette of egg beaten up in milk has been recommended; such a preparation, however, is liable to adhere to the snake's head and obstruct the nostrils.

A list of the main food preferences of some common genera of snakes is given in Table elsewhere in this chapter.

Water

Reptiles should always have water available for drinking and bathing. Some species such as the heloderms, which live in dry regions, possibly absorb moisture through the skin and in captivity spend much time lying in their bath. Chamaeleons seldom drink from a dish but will take water from foliage that has been sprayed.

Anaesthesia

Ether is convenient as a general anaesthetic, the animal being placed in a closed glass container and quickly removed as soon as it becomes relaxed and motionless. Swabs of cotton wool soaked in ether can be applied to the nose if there is any tendency for the animal to come round during operation.

It is, however, easy to kill reptiles in this way, and more refined techniques involving the use of barbiturates are recommended.

Table 7.1: Food Preferences.

Family	*Genus*	*Species*	*Food (common names)*
BOIDAE	Boas and pythons	Birds and mammals	
COLUBRIDAE	*Natrix*	E.g. grass snake	Amphibians and fish
	Thamnophis	Garter snakes	Amphibians, fish, earth worms, chopped meat
	Heterodon	Hog-nosed snakes	Toads
	Zamenis	E.g. dark-green snake	Lizards, rodents
	Coluber	E.g. aesculapian and four-lined snakes, American racers	Lizards, rodents
	Ptyas	Rat snakes	Rodents
	Pituophis	Bull and pine snakes	Rodents, birds
	Coronella	Smooth snakes	Lizards, young rodents
	Lampropeltis	King snakes	Reptiles, rodents
	Dasypeltis	Egg-eating snake	Eggs only
	All colubrine tree snakes		Lizards, frogs, small birds
ELAPIDAE	*Naja*	Cobras	Frogs, birds, rodents
	Dendroaspis	Mambas	Rodents, birds
	Bungarus	Kraits	Frogs, rodents
VIPERIDAE		Vipers, rattlesnakes etc.	Small birds, mammals; fish and frogs for aquatic species

Euthanasia

Most reptiles can be killed easily with ether or chloroform or by decapitation, though the latter may cause the body to go into violent spasms which persist for some time. Tortoises are very resistant to inhaled anaesthetics.

C. C. D. Shute recommends the following method of preparing tortoises for class dissection. Two drops of pure nicotine are placed in the mouth, the animal being rendered unconscious almost immediately. The plastron can then be removed with a circular saw.

The heart will still be beating slowly. Lead chromate suspension is then injected into the ventricle, the animal being fixed for a fortnight or so in formalin and then preserved in alcohol to which a little glycerine has been added. Nicotine is highly toxic to humans as well as reptiles and should be used with the greatest caution.

Experimental Techniques

Techniques for operating on reptiles, collecting their body fluids, and so on are still in their infancy, apart from methods of venom-extraction, which have been described in detail for rattlesnakes by Klauber (1956). Bragdon (1953) gives a useful account of various surgical techniques which can be applied to snakes.

In experiments on the brain, parts of the skull can be removed with a dental drill; the wound can be subsequently closed with gelatin sponge sealed with 5 per cent celloidin or with a solution of Perspex in acetone. Incisions in the skin and soft tissues can be closed with sutures.

The introduction of penicillin and sulpha-drug crystals may help to prevent infection. Techniques used in temperature studies are given by Benedict (1932) and Bogert and del Campo (1956) and in studies on learning by Boycott and Guillery (1962).

Transport

For journeys of up to a few days, small reptiles can be placed in boxes or tins with ventilation holes, packed fairly tightly with moss or hay which is *very slightly* damp. It is important to avoid overcrowding as reptiles may injure each other with their claws.

Snakes will also travel conveniently in bags which are permeable to air; several bags, each containing one or two snakes, may be packed close together in a single wooden box.

Larger reptiles such as monitor lizards, crocodilians and tortoises should if possible be packed individually in separate containers or one partitioned container. On long journeys they will need water and should be hosed down at frequent intervals, the water being introduced through a hole at the top of the crate and allowed to drain away through holes in the bottom.

Except perhaps in the case of small lizards, feeding is unnecessary. Tropical reptiles in transit should not be exposed to temperatures of much under 20°C (68°F) for more than a few hours; journeys by air in unheated freight compartments may be fatal, and permission should be sought from airline companies for the reptiles to enjoy passenger conditions of temperature and pressure.

Reptiles in transit may also be in danger from overheating, and their containers should never be left in strong sun. Special precautions should be taken over the transport of poisonous snakes, and all containers should be thoroughly checked for holes through which even young snakes could pass.

Many poisonous snakes (e.g. big vipers and rattlesnakes) are viviparous, and young may be born during a journey. It should be noted that the very long fangs of some viperine snakes can easily penetrate the cloth of bags or ventilation holes in box containers; the latter should therefore be covered with wire gauze. Accurate and conspicuous labelling of boxes containing venomous reptiles is, of course, essential.

INJURIES AND DISEASES

Only some of the more important pathological conditions likely to occur among laboratory reptiles will be mentioned here.

Treatment

Wounds should be washed and treated with antiseptic liquids in the case of land reptiles, or with powders in the case of aquatic ones. If extensive they should be stitched. Sulpha drugs and penicillin should be used in the treatment of severe wounds.

Abscesses and cysts should be incised and packed with suitable preparations. Snakes are very susceptible to infections of the mouth and jaw bones, which may result from injuries or striking the nose against the cage.

If taken in time such conditions can usually be cured by gently removing the dead tissue and pus, and powdering the mouth with sulphanilimide or penicillin powder, or dressing with chloromycetin in glycol. Preliminary treatment of the infected area with iodine has been recommended.

Diseases due to Unfavourable Conditions of Life

Persistent refusal to feed and eventual death through starvation may be caused by keeping reptiles at temperatures which are too low, or denying them access to sufficient sunlight. Diseases resembling rickets may be arrested by allowing greater exposure to sunlight and by giving cod-liver-oil preparations and supplementary calcium.

Blisters often develop beneath the superficial layers of the scales of snakes kept under too damp conditions; they clear up when the conditions are remedied.

Ophthalmia

Reptiles, especially tortoises emerging from hibernation, often develop swollen and inflamed eyes which should be treated with drops of some silver preparation such as 5 per cent protargol, argyrol and Procid.

External Parasites

Wild lizards, snakes and tortoises are often infested with ticks. These can be killed by painting with paraffin or methylated spirits, and then should be picked off with forceps, care being taken not to leave the head in the host's body.

Squamata, especially snakes in poor health, are frequently attacked by mites such as *Ophionyssus spp.* which may carry pathogenic bacteria capable of producing a fatal disease very quickly. Mite infection is transmitted from reptile to reptile and, since the mites have good powers of locomotion, can travel from cage to cage.

It is important to recognize early a mite-infested animal so that it can be segregated and the condition be prevented from spreading through the collection. In snakes the first signs of infestation are usually the appearance of silvery-white deposits of mites' faeces on the scales.

A few fully grown females may be seen appearing as shiny red or black beads up to about 1 mm in size. Immature mites are like specks of fast-moving dust and may be detected on the hands after handling an infested snake.

Mites can be eliminated by repeatedly washing the entire animal in warm water and wiping it down with cotton-wool. It is best to do this in a special bath outside the cage; all mites washed off must be removed from the bath before another reptile is immersed.

The use of various other preparations have been recommended but DDT should not be used, as it is very toxic to reptiles. An infested cage should be thoroughly cleaned with boiling water, or with a strong disinfectant such as lysol followed by soapy water. No reptiles should be placed in it for at. least a week.

Mite-infestation is favoured by dirty cages containing pieces of sloughed skin etc. In clearing out infested cages or handling infested reptiles care should be taken to prevent the mites from spreading around the neighbourhood.

Diseases due to Internal Parasites and Micro-organisms

All reptiles are subject to nematodes. Tortoises have been cured by the administration of 1 gr (0.65 mg) santonin with 1 gr calomel, followed up 24 hours later with a teaspoonful of castor oil.

The treatment, without the calomel, should be continued once a week for six weeks to ensure the removal of any young worms subsequently hatched.

Reptiles are also subject to many other types of infection, such as amoebiasis, which causes a fatal enteritis, and pentostomiasis (infection by arachnid parasites).

For the treatment of amoebiasis some success has been achieved with the drug Entamide. Some observations on internal parasites of reptiles are given in Smith (1954), and on reptile diseases in general in Reichenbach-Klinke and Elkan (1965).

8

The American Cockroach

The Cockroach is the insect most commonly used as a type animal in the teaching of biology. *Periplaneta americana* has some advantages over *Blatta orientalis,* the other species often used for this purpose, in having larger wings which are present in both sexes, and a less offensive smell, and last but not least it is not likely to become a serious pest if it escapes from captivity.

Cockroaches are also used extensively in the testing of insecticides. *P. Americana* has recently become more important with the development of insecticide resistance in the German cockroach, *Blattella germanica;* though resistance has been known in the U.S.A. for some years, the first definite case was not reported in the United Kingdom until 1960.

The American cockroach is comparatively rare in Britain although common in ships. It is however introduced, together with other species, with banana cargoes.

Cockroaches are also used for general experimental work but attention should be drawn to considerable variability in their biology and life history. Little is known about their genetics, although Jefferson (1958) records a white-eyed mutant from a mine in South Wales.

The Blattaria as a group offer a rich source for research since few of the 3,500 described species are known in detail.

Cockroaches have been used as live food for various animals, e.g. mantids, scorpions and amphibians. Washburn (1913) described a trap for cockroaches.

GENERAL BIOLOGY

In warm conditions the American cockroach is an active animal and can run fast even up a vertical surface. The claws are used if the material is rough, and the plantulae or adhesive pads at the tips of the tarsal segments if the material is smooth, e.g. in the case of glass. On an inverted smooth surface it uses an adhesive lobe, the arolium, which lies between the claws.

The young nymphs are unable to climb the vertical sides of a clean glass vessel. Willis *et al.* (1958) state that this ability is acquired after the sixth or seventh moult. The cockroach does not fly in captivity, though the wings may be used in the tropics and may be flapped in excitement or when the male displays before mating.

Cockroaches have a tendency towards cannibalism which may become serious if adequate food is not provided. They also eat their own eggs or may damage the capsules so that the eggs lose water and do not develop.

Adults are apt to fight or bite each other, and so in a large colony there are often some animals with one or more legs severed. Cockroaches are gregarious; Gould and Deay (1938) found that the nymphs thrived much better when several were reared in the same container rather than singly.

Although in warm climates the American cockroach is found in sewers, it regularly cleans itself with the mouth-parts and legs. There are no grounds for assuming that it should not be kept, like other laboratory animals, in hygienic conditions.

A well-marked diurnal locomotory rhythm has been examined by Cloudsley-Thompson (1953), who found that it is correlated with alternating light and darkness but not with fluctuating temperature and humidity, although constant temperatures have a depressing effect. At room temperature and humidity and in natural daylight and darkness, the animals are active at night and quiet during the day.

REPRODUCTION AND LIFE HISTORY

The adult males may be distinguished by their anal styles and the females by the boatshaped end to the abdomen. In the newly-emerged females the oocytes are relatively small and do not reach full size for a week or more.

The males are attracted by the odour of a cuticular grease on the virgin females and become excited, flapping their wings. When contact is made, the male turns and copulates in the opposed position, i.e. the heads face in opposite directions; this lasts an hour or more.

The spermatozoa are transferred in a spermatophore and received by the spermatheca of the female where they may remain viable for a considerable time. The percentage of viable eggs decreases however with age if males are absent.

The eggs are laid in egg-capsules or oothecae, the walls of which are produced by the colleterial glands. Normally each capsule contains sixteen eggs in two rows of eight but the number may be reduced by some abnormality in the ovaries; on the other hand the number is sometimes greater.

The dorsal edge or keel of the capsule is modified into a series of respiratory chambers which admit air to the developing embryo. The capsule is carried by the female at the end of the abdomen for a few hours or a day or, in cold weather, for several days.

It is either buried or attached by a glutinous substance to some object in such a way that the keel is free. Internal pressure exerted by the fully formed nymphs causes the capsule to split along the keel, but not all nymphs may emerge because some eggs may have been placed the wrong way up in the capsule, and some nymphs may be trapped when it closes after the first young have emerged.

The nymphs are at first covered by their embryonic membranes which sometimes remain caught between the two sides of the keel. Normally the quickest method of distinguishing between capsules which are empty and those which contain eggs is to roll them lightly between the ends of the thumb and first finger.

The nymphs are similar to the adults but lack wings. On emergence, and after each moult, the body is white, but it acquires the normal brown colour during the next few hours. Authors differ as to the number of moults.

Gould and Deay (1938) give 13, Gier (1947) 10; while in 13 insects reared in isolation Willis *et at.* (1958) found between 9 and 13 moults. Styles are present in both sexes in the nymphs but the females can be distinguished by a median notch at the end of the ninth sternum, which is absent in the male, or present only to a small degree.

Development is greatly influenced by temperature. Gould (1941) obtained an average of 520 days for full development at about 25·5°C (78°F) and 195 days when the mean was 28·3°C (83°F). Similarly the mean incubation period varied from 58·5 days at 24·6°C (76°F) to 48 days at 26·0°C (79°F) and 34·5 days at 29·2°C (85°F).

Table 8.1: Comaprison of records of biology of the American Cockroach.

Source	*Adulthood (days)*				*Preovi-position period (days)*	*Days between egg cases*	*Egg cases per female*	*Days of incubation*
	Female		*Male*					
	Av.	*Max.*	*Av.*	*Max.*				
Gould and Deay	438	1588	—	—	13·4	5·9	58·8	55·3
Griffiths and Tauber	225	706	200	362	20·0	6·8	21·1	53·0
Rau	310	330	82	175	—	7·4	11·9	43·0

Some data on the biology of the animal are given in Table elsewhere in this chapter.

ACCOMMODATION

Humidity

This is one of the most important factors in the keeping of cockroaches and the following points should be noted:

(1) The egg-capsules can develop and hatch in very low humidities provided they have not been damaged in any way. The capsules are particularly susceptible for the first three days after formation; after this a high percentage of eggs may be hatched from capsules with keels removed if the relative humidity is 90 per cent. One of the principal functions of the capsule is to prevent loss of water.

(2) The first instar is also very susceptible to drying and should be kept at 70 per cent RH for the first week.

(3) Older nymphs and adults can live quite healthily in much lower humidities, provided they have access to water.

(4) Humidities above 60 per cent may lead to outbreaks of mites or Psocids.

(5) With careful sterilization of containers and food, and exclusion of crawling animals by means of an oil barrier, self-contained colonies can be maintained successfully in RH 70.

(6) At RH 75 moulds may infect both food and egg-capsules. Humidity can be controlled in several ways:

 (a) Warm air from a heating element may be blown over wood wool kept moist with water. A humidistat controls a heater and fan for this purpose and a thermostat controls a second heater which regulates the temperature of the breeding-room.

 (b) A humidistat may be connected to an aerosol humidifier.

 (c) Air may be circulated over saturated salt solutions. A salt is chosen which, in a saturated solution, gives the vapour pressure corresponding to the humidity required at the correct temperature.

Temperature

For breeding, a minimum of about 27·7°C (82°F) is required, but the animals can be maintained at a lower temperature (24 to 25°C, 75 to 77°F) without difficulty. They become almost inactive at I0°C (50°F), and near freezing-point they appear to be dead but may be brought round if not subjected to this temperature for too long, e.g. about 2 or 3 hours.

Preventing Escape

Since cockroaches are active and also potential pests, special precautions should be taken to prevent escape. The following methods may be used:

(1) The cage should have a well-fitting lid. Care must be taken with wooden lids enclosing lamps or heating elements, as even adults may escape when only slight warping has taken place.

(2) An electric fence round the top of the chamber. This can be made from two strips of copper foil firmly tacked to a wooden frame and differing in electric potential by about 35 V. Each of the strips should be not less than 1 cm wide and they should be placed close together, e.g. 2 or 3 mm apart; if space permits it is better to increase the width of each strip to 2 cm. If the fence is facing downwards the animals will touch it with their antennae before they walk on it and so be spared a shock.

(3) Petroleum jelly on glass. This should be applied to form a barrier 8 cm wide on a vertical, or better still a downwardly-facing, surface. It is suitable for small containers but is not entirely reliable and will collect dirt.

(4) An oil barrier around the chamber. A suitable grade of oil which is colourless and odourless is Technical White Oil. The barrier should be not less than about 0·5 cm deep and at least 8 cm wide. Dust which falls on this oil sinks to the bottom, so that the surface remains clean.

(5) P.T.F.E. fluon dispersion on glass.

Nutrition

Cockroaches can live on very poor diets. Gier (1947) found that growth was retarded when the American cockroach was reared aseptically; but it is not clear what part, if any, micro-organisms normally present in the digestive system play in supplying necessary constituents to the diet.

Cockroaches can live healthily on fresh food, such as potato and apple if these supply enough water but not on dry food alone, especially if the conditions are dry. Water is always important. Willis and Lewis (1957) found that at R.H. 36 they could live as long with no food or water as they could with dry food alone, whereas they lived much longer with water alone. A watering fountain consisting of a full beaker inverted on a Petri dish with cotton-wool to prevent young nymphs from drowning is suitable for most purposes.

The water should be changed when it becomes fouled by saliva. A test-tube plugged with cotton-wool lying on its side is useful where space is limited. Noland (1956) describes a watering device used in rearing-tubes.

Dry foods have the advantage that they can be sterilized by heat but the disadvantage that they may be strewn all over the cage. Many forms and mixtures have been used, such as dog biscuits and rat cubes.

The medium used at the Pest Infestation Laboratory, Slough, contains rolled oats, wheatfeed, fishmeal, and dried yeast powder in the ratio 9:9:1:1. After moulting, cockroaches consume their own exuviae, but not the legs.

BREEDING

The Tolworth System

The following method is used at the Infestation Control Laboratory of the Ministry of Agriculture, Fisheries and Food at Tolworth, Surbiton, Surrey. The colonies are kept in larvae cages 40 cm high and 20 cm in diameter with plastic sides, metal bases, and perforated metal lids.

Most of each cage is occupied by 10 platforms of hardboard measuring 15 × 10 cm, separated by 1.8-cm wooden blocks and held together by two pieces of wire which pass through them; these also provide two handles at the top.

Water and food are placed on the top platform. The colonies are subcultured every three months. A hole in the plastic, normally stoppered with a rubber bung, provides an inlet for carbon dioxide which is used to anaesthetize the cockroaches before the lids are removed.

The cages rest on raised stands on galvanized iron trays which contain a layer of Technical White Oil. This provides a reliable barrier to cross-infestation from other insects and mites.

The humidity in the cages appears to be the same as that in the room, which is kept at an R.H. of 70 per cent and a temperature of 27.7°C (82°F). The method is particularly suited to keeping a large number of isolated colonies of insects in one room, and is used for Blatta orientaiis, Blattella germanica and Periplaneta australasiae.

The Slough System

After the first week the adults and nymphs are kept at R.H. 30 but this does not appear to retard development provided there is ready access to water. The egg capsules are buried in the dried food mixture which is contained in a crystallizing dish; less than 2 per cent of the egg capsules are damaged under these circumstances.

They are collected weekly by passing the whole of the food medium through a sieve; they are then washed, dried and incubated in another room at R.H. 70.

Breeding Observation Chamber

The main features of this method are the false floor of 0·6-cm wire mesh which induces the females to lay the majority of their egg capsules in holes in a piece of hardwood. The latter, being under the hanging boards which provide space for the main part of the colony, is shaded from the lamps, but room is left to move the boards with their stand when the egg capsules are removed.

These are placed in a separate small chamber, not shown, with a small shallow tin containing water and fresh food. The newly emerged nymphs fall into the tin and find more humid conditions under the food (sections of apple or potato) or in a test-tube half filled with water, plugged at this level with cotton-wool, and left on its side.

The nymphs are removed once or twice a week and put into another nymphchamber with small platforms; this is a small version of the arrangement used in the Tolworth system. The nymphs are put into the main chamber when they are larger.

The colony can maintain itself successfully with less attention if the piece of hardwood is removed so that all the egg capsules fall through the wire mesh. It is important in this case to ensure that the adults cannot get into the trays.

The egg capsules may either be collected at less frequent intervals before they have hatched, sieved from the droppings, and put into a separate container, or be allowed to hatch in the trays; this may lead to some mortality but a sufficient number survive to maintain the stock.

Care should be taken in both cases not to allow the pile of egg capsules to become too deep or the ones underneath will be attacked by moulds. The advantages of this breeding observation chamber are that the egg capsules are protected, the adults and larger nymphs do not come into contact with their droppings, the chamber is self-contained although a large colony can be kept, all stages are visible, and the temperature is variable.

There is a temperature gradient of about 3°C (5°F) between the top and bottom of the boards in the shade. In practice a range of 23 to

26°C (73 to 79°F) has proved satisfactory. Under the lamps the temperature may rise above 30°C (86°F). The humidity is rather less than that of the room.

If fresh food is used, such as apple, potato, or carrot, the only debris which collects in the trays other than droppings are the legs not consumed after moults and to some extent wings detached after damage. Animals which have had legs bitten off can quickly be spotted and removed as they cannot cling to the boards but lie on the wire.

Cummings and Menn (1959) describe a cage in which the wire mesh becomes the floor and the egg capsules are collected in a tray below. A water and feeding device is shown, built into one side so that the lid does not need to be opened regularly.

Piquett and Fales (1952) recommend that a space of 5 cm' be allowed for each adult animal, but the glass sides and parts exposed to the lamps should not be included in estimating potential capacity of the chamber.

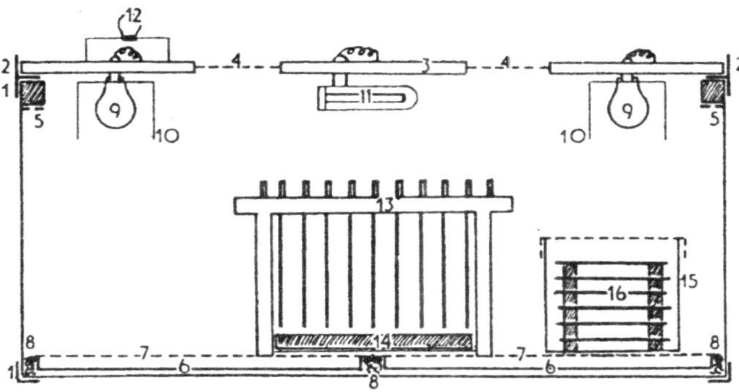

Figure 8.1: Breeding observation chamber.

HANDLING, CONTROL AND KILLING

It is difficult to pick up individual active cockroaches by hand without damage, but a long pair of fine forceps can be quite satisfactory if the insect is held in the middle of the abdomen.

Larger numbers should first be subdued either with carbon dioxide or by allowing the chamber or a separate container to become chilled, or by a combination of both methods.

Carbon dioxide is more suitable for rapid handling, but the animals should not be left in the gas for more than a few minutes. They are killed in chloroform vapour.

TRANSPORT

Exposure to the cold is, perhaps, the difficulty which is most likely to be encountered, but in normal summer temperatures this should present no problem, even for a long journey. Egg capsules should be packed with cotton-wool and the container sealed.

Nymphs and adults should be provided with something on which to rest, such as crumpled paper, and put into a tin with a perforated lid. If the animals are accompanied on a long journey, it may be more convenient to put them into a large glass bottle.

Provided the paper does not reach the neck, the top can be removed without much difficulty for feeding. If a warm radiator is available, the bottle may be placed close to it and animals will find their own optimum between the hot and the cold sides.

There is no difficulty about sending the animals by rail. Regulations prohibit the sending of insects, other than certain ones specified, by post but in Britain provision is made for individual cases on application to the G.P.O. Headquarters Building, St Martin's-le-Grand, London, E.C. 1. British Overseas Airways Corporation will undertake to bring cockroaches from abroad and as far as possible at the temperature specified; they will also arrange customs clearance at London Airport and deliver by van in the London area, or dispatch by rail or air to other parts of the country.

PARASITES

Attention has already been drawn to mites. Though most of these are harmless in small numbers, they may overrun a colony. The species *Pimeliaphilus podapolipophagus* is however a true parasite. Fisk (1951) eliminated mites, possibly of this species, in a cockroach colony, using a 5-per-cent spray and a 5-per-cent dust of *p*-chlorophenyl, *p*-chlorobenzene sulfonate. Nematodes may be present in large numbers without apparently affecting either health or fertility.

Egg capsules which have been sent from abroad should be examined for signs of the eggs of parasitic wasps, especially Evaniidae. Griffiths and Tauber (1942b) dipped the egg capsules in 70-per-cent alcohol for 10 seconds to prevent fungal growth.

9

BATS

The use of Bats as experimental animals has been unduly neglected in view of the unusual problems of their physiology, behaviour and flight. It must be admitted that it is often difficult to provide the diet and reasonably natural conditions demanded by these highly specialized animals in captivity.

For this reason most Chiroptera are not suited to normal methods of animal-house management and require strictly personal attention. Nevertheless they are normally long-lived, and with care many species can be maintained in apparent health for several years.

The megachiroptera are generally responsive to human attention and make attractive pets when tame. This is less obviously true of the microchiroptera, but probably only because their behaviour is largely non-visual; most soon learn to associate the hand with food and will then seldom bite.

Comprehensive accounts of the general biology of bats are given by Allen (1939), Eisentraut (1957), Griffin (1958) and Ryberg (1947). Information on the care of several insectivorous species is given by Gates (1936), Mohres (1951) and Orr (1958). Kulzer (1958) describes in detail the maintenance and breeding of a megachiropteran, *Rousettus aegyptiacus*. The general principles described by these authors may be applied throughout the order since the requirements of other species differ mainly in their diet.

ACCOMMODATION

Galvanized iron cages without a raised floor grid are convenient for the larger megachiroptera but metals are better avoided for smaller bats, since urine may form a noxious solution and be reingested. Wood and hardboard are the best materials, although celluloid, perspex (plexiglass) and glass have been used successfully.

All joints and doors should fit tightly to prevent escapes, as most bats can squeeze through surprisingly small cracks. Many species are sensitive to draughts, and over-ventilation should be avoided. Solid walls with one small window covered by plastic fly-netting form a good arrangement.

For temporary or even permanent quarters, a glass aquarium can be fitted with a partly perforated lid and is ideal for observation. Bats may safely be confined in cardboard cartons or light cloth bags for transportation, as they will not attempt to gnaw their way out.

Although most bats are gregarious, different species should not be kept together in the same cage. If this must be done, great care should be taken in the choice of cage-mates.

For fruit bats, vampires and other forms that produce soft droppings, the floor of the cage may be covered with sawdust and cleaned daily. Insectivorous bats produce dry faeces so that a sheet of newspaper is quite sufficient.

The species that descend to the floor of the cage dislike sawdust and are frightened by the crackling noise of paper. A bare floor is best in these cases and is easily swept clean.

Perches for large fruit bats are best made of wood dowel, as continued suspension from wire bars can lead to excessive curvature of the claws. Horizontal bars on the walls should be so spaced that descent to the food is easy.

For smaller bats the cage walls may be roughened, or lined with plastic netting, to give a suitable purchase for climbing. When roosting, rhinolophids and some other bats hang freely from a horizontal surface and for these forms the roof of the cage also should be lined. Other species perch on vertical surfaces; vespertilionids and molossids normally seek crevices, and like to sleep with solid surfaces both dorsally and ventrally.

A suitable roosting-box for these forms has been devised by the Earl of Cranbrook (personal communication) and one form is shown in Figure elsewhere in this chapter.

The back may be slatted with slivers of wood or lined with plastic netting. The glass front must be well stoned to remove sharp edges and is held in place by an elastic band at the top. The chamber is placed against one wall of the cage and is always adopted by vespertilionids,

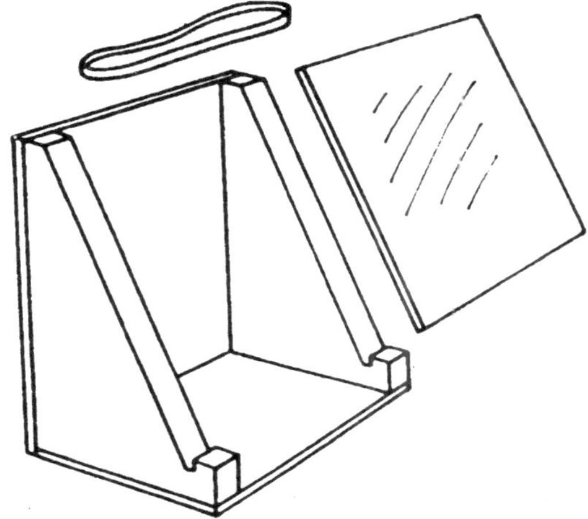

Figure 9.1: A roosting chamber, about 12 cm high, suitable for most Vespertilionids and other crevice-seeking bats.

which crawl up inside as far as possible, small bats going higher than large ones.

These chambers are easily cleaned and facilitate observation, removal, and handling of the animals, which readily return to them, even outside the cage.

Lord Cranbrook has shown that bats are not disturbed by the light and show no preference for dark conditions when these are available, although De Coursey and De Coursey (1964) showed that periodicity of light is an important factor in regulating daily cycles of activity.

Temperature

Eisentraut (1934), Burbank and Young (1934) and Reeder and Cowles (1951) demonstrated that bats of temperate latitudes are poikilothermic, at least when inactive. During daily sleep, as well as in hibernation, the peripheral body-temperature falls to a little above the ambient.

They can therefore be kept at a wide range of room-temperatures, but need greater amounts of food when in warm conditions. Torpid bats can be awakened by handling, but must be allowed to shiver for several minutes to raise their body-temperature before they can become fully active.

They can also awake voluntarily and will soon learn to do so at any time of the day before a regular feeding-time. Tropical microchiroptera are sensitive to low temperatures and must be kept warm if transported to higher latitudes. Megachiroptera are homoiothermic, shivering when cold, and must be kept at a temperature of 22°C or higher.

Exercise

Many bats can be kept in apparent health with little or no exercise, but for others a daily flight period is essential. It is seldom possible to provide the larger megachiroptera with suitable free flight conditions. *Pteropus giganteus* has a wing-span of 1 m and is unable to turn in a small radius.

Nevertheless it can be kept satisfactorily without any exercise in permanent quarters less than 1 m cube. Some swift, high-flying vespertilionids (e.g. *Nyctalus) and* molossids show no inclination to fly in normal-sized rooms, although the ability is not lost even after several months of captivity.

Other species benefit from occasional flights but are easily lost if released in a furnished room, which always offers a multitude of hiding-places. Ideally their cage should be kept in an empty, well-sealed room and left open each evening.

If this is not possible, the cage should either be large enough to permit short flights or so small as to inhibit flight completely. Attempts to fly in a cage which is not quite large enough lead to self-inflicted injury and deterioration of health. Nectar-feeding bats such as *Glossophaga* can exercise adequately in quite modest cages in virtue of their slow, hovering flight.

Rhinolophids deprived of exercise develop swollen wrist joints in a few days and make frantic attempts to escape from small cages. Mohres recommends that a cage for these animals should measure at least 70 × 60 × 50 cm and that they be allowed free flight each night.

This does not present such a problem as for vespertilionids, since they hang freely and are easily located if they fail to return to their cage. It is rather important, however, that no major changes be made in the arrangement of furnishings, particularly in the position of the cage, when once the bats have become accustomed to them.

All bats appear to have a highly-developed topographical memory which overrides their acoustic orientation in familiar surroundings. Changes can therefore lead to collisions, and a transposed cage, once lost, may take some time to be rediscovered.

HANDLING VAMPIRES

Even when so gorged with food as to be unable to fly, vampires can run and jump with surprising agility and speed. Once free they are very difficult to recapture and should therefore be handled with great care.

Thick gloves should always be worn as the large sharp incisors are specialized for producing deep wounds and there is a real risk of rabies infection. The best way to seize an untamed vampire is by one leg with blunt, long-handled forceps. The animal may be grasped firmly with the other hand before the forceps are loosened.

Nutrition

Food varies a great deal throughout the order and can best be determined for each species from accounts of its general biology and habits. Goodwin and Greenhall (1961) list the food of over 60 species from all the major neotropical groups (those species found in Trinidad), many of which have unusual or mixed diets. Phillips (1922) and Rosevear (1965) give notes on several families from the Old World tropics.

Vitamins

All diets should be supplemented with vitamin preparations. Mohres maintains that oily preparations are too difficult to administer in amounts suitable for very small mammals, and excesses often cause upsets leading to vomiting.

Orr (1958) used Stuart Formula Liquid, a comprehensive but completely water-soluble preparation manufactured by the Stuart Company of Pasadena, California. This has been used with success for bats and other small mammals in the U.S.A., Germany, and Britain. It contains vitamins A, B_1, B_2, B_s, niacin and niacinamide, D, E, and d-panthenol, together with iron, iodine, manganese, sugar, sodium benzoate and malt (vitamin B complex).

There is some evidence that preparations containing an excess of vitamin A can cause reversible depilation in both megachiroptera and microchiroptera. This is unsightly, but there do not seem to be any other adverse effects. After some time in captivity on unsupplemented diets many bats are prone to weakening of the jaws and limbs, and general listlessness. Stuart Formula Liquid effects a complete cure in some cases.

INSECTS AND MASH

All temperate and a large number of tropical microchiroptera are purely insectivorous. Nearly all forms will accept and thrive on mealworms *(Tenebrio* larvae) which are easily cultured. Maggots (Calliphoridae) and early instars of many orthoptera are readily taken also.

Gates (1936) devised an artificial mash which has proved effective in several modified forms. This consists basically of bread or cracker biscuits, hard-boiled egg, cream cheese, milk and banana. Meat, insects, vegetables, yeast, malted milk and other items may be added, and the whole is chopped and mixed to a fine crumbly texture.

Gates remarks that all bats particularly like cream cheese and banana. Since mealworms themselves have a purely farinaceous diet, they may lack certain constituents necessary to the well-being of bats.

This deficiency can be corrected in part by occasional substitution of orthoptera fed on green stuffs, but it is safer to provide known additives in the form of vitamin preparations. Buchholz (1958) has investigated the digestive enzymes of two vespertilionid species.

Presentation of Food

Newly captured insectivorous bats show variable readiness to take food when not flying; some will eat a large meal within minutes of capture, while others must be coaxed into taking the first bite. Obstinate individuals may be tempted by smearing crushed mealworms on their lips or directly provoked to snap at a persistently proffered larva.

Once the first mealworm has been tasted there is seldom any reluctance to accept further specimens. Although at first they must be fed by hand, or even in the hand, most individuals learn within a few days to take their own food.

Nevertheless eating-habits are not completely flexible, for many vespertilionids continue to tuck each insect against the interfemoral membrane as in flight. This token movement serves no useful purpose on the ground and often results in a complete somersault.

Live mealworms may be placed in dishes in the cages of vespertilionids and other crawling bats, which awaken and eat when hungry. But, as Orr points out, most bats over-eat if given the opportunity, and it is difficult to ensure that several bats in the same cage get equal rations, since the first to awake takes the major portion.

This may be prevented if individual portions are put in separate deep dishes on the bench and the bats are placed in these under glass covers. Each animal wakes rapidly and after finishing its meal may be returned to the cage.

This procedure is well worth the slight extra labour involved, since it allows the consumption of every individual to be observed and controlled. All bats have large appetites and when active need the equivalent of one to two large mealworms per gram of body-weight per day. The smaller bats of course, prefer smaller mealworms as they find large ones difficult to chew.

Rhinolophid bats and other hanging forms are best fed by hand, but may learn to eat from a dish so placed that they can reach it while perching. They can seldom be persuaded to eat when placed in a dish themselves.

Fruit

All the megachiroptera and a number of tropical microchiroptera are purely frugivorous. Other microchiroptera normally eat both insects and fruit, though some are also partly carnivorous, and may survive well on fruit alone, which is easier to provide and to dispense than insects. The food preferences of some neotropical fruit bats have been analysed by Greenhall (1957).

In general, soft, sweet fruits are preferred, for example bananas, mangoes, peaches, plums and melons. Grapefruit and sweet oranges may be taken, but citrus fruits are generally rejected if alternatives aie available. A diet of bananas with Stuart Formula Liquid appears to be adequate for megachiroptera.

Should exotic fruit be in short supply in temperate latitudes, sweet apples alone will suffice for a time. The larger megachiroptera are able to peel bananas by clasping the fruit to the chest with one foot and tearing the skin with their teeth. Apparently whole citrus fruits are not sufficiently attractive to merit more than cursory treatment.

If such less attractive food must be included, it is advisable to dice all the components and give a well-mixed fruit salad. Smaller frugivorous bats can only tackle whole fruit if it is overripe and very soft; they puncture the skin or bite a small hole to suck at the pulp.

Flesh

Several of the larger microchiroptera sometimes take small birds and mammals (Dunn, 1933) and at least two species, *Vampyrum spectrum* of the New World and *Megaderma lyra* of the Old World, are completely carnivorous. Ditmars (1935), Gleadow (1906), Green (1906), Phillips (1922), and Primrose (1906) state that these bats capture and eat frogs, birds, small bats and rodents.

Captive *Phylloderma* accept geckos and the closely related *Trachops* is said to hunt and capture these reptiles.

Many insectivorous bats will occasionally accept small pieces of raw meat or liver as a substitute food.

Blood

The true vampires of the neotropics (*Desmodus* and *Diaemus*) feed exclusively on the blood of mammals and birds respectively. The feeding habits of *Desmodus* have been described by Ditmars and Greenhall (1935), Trapido (1946), and Goodwin and Greenhall (1961), who all report that it is very hardy and can easily be kept in captivity on a diet of defibrinated blood.

Each animal weighs about 30 g and may consume 50 g of blood per day from a dish, although 10 to 15 g is usually sufficient. The author finds that vampires readily learn to drink from drinking tubes, which keep the blood in good condition and improve cage hygiene.

Blood is taken by a lapping motion and passes along a ventrolateral groove on either side of the cylindrical tongue. *Diaemus* will take small quantities of ox blood but does not survive for long unless avian blood can be provided. After feeding, vampires are often incapable of flight and *Desmodus* well deserves the specific name *rotundus*.

Fish

Several microchiroptera, including *Noctilio* (Noctilionidae) and *Pisonvx* (Vespertilionidae) are specialized for a diet of fishes, which they seize with their large feet when flying over the surface of the sea or fresh water.

It is not yet clear how the fish are located or whether random sampling is involved. The author helped D. R. Griffin to maintain a number of *Noctilio leporinus* for several weeks in Trinidad. These large, robust bats were kept in an outdoor flight cage, which included a small pool.

At night they eagerly seized pieces of fish from forceps or from the surface of the pool. They also caught and consumed insects which entered the cage, especially if these fell into the water. It would no doubt be possible to keep this species, and possibly some of the others, under such conditions for a considerable time.

Nectar and Fruit-Juices

Members of the sub-family Glossophaginae (family Phyllostomidae) of Central America feed on the nectar of night-flowering trees and shrubs and upon the juices of overripe fruit and possibly a few insects. They have extremely agile, hovering flight and very long protrusible tongues, thus forming a nocturnal counterpart to the humming birds.

In captivity several species accept a watery suspension of mashed banana or mango in strong sugar solution. *Glossophaga* will thrive on this diet for many weeks. Despite its close affinity with the insectivorous and carnivorous *Phyllostoruus hastatus, Phvllostomus discolor* is best treated as a nectar-feeder in captivity.

Very small megachiroptera such as *Megalot'lossus* and *Nauonvcteris* also need a fluid, pulpy diet.

WATER

All bats, including the nectar-feeders, need a plentiful supply of clean drinking-water but do not seem to use drinking-tubes. Water is best supplied in shallow dishes, which must be cleaned each day to remove droppings and other contamination.

For smaller bats the depth should not exceed 3 to 5 mm, since there is a danger that sick or sleeping animals may drown in it. Vespertilionids and other crawling forms approach the dish along the floor and lap over the edge.

Pendant forms such as rhinolophids and the megachiroptera are reluctant to cross the floor, so the dishes must be placed where they can be reached from a wall or the roof of the cage.

Most bats can tolerate wide ranges of humidity if plenty of drinking-water is available. However, temperate microchiroptera at low temperatures may remain torpid for long periods and must then be maintained at high humidity to minimize water-loss.

They benefit if they and their cages are sprayed frequently with a mist of clean water. Megachiroptera may also develop wing sores in dry conditions, especially if the cages are too small for complete preening; this action requires a lot of space in the larger species.

These animals may also be sprayed daily with water, and a wet sack placed in their cage helps to keep the wing membranes moist. The dry conditions may not appear where several animals are kept together.

BREEDING

Several species of megachiroptera will breed readily in captivity and colonies may be maintained over several generations. Mating, parturition, suckling and weaning have all been observed. The gestation period of *Pteropus giganteus* is about 180 days.

Many species of microchiroptera have given birth to young in captivity and in several cases these have been successfully reared. There are few certain accounts of mating in captivity which later led to successful birth, although vampires can be bred easily.

All stock must therefore be captured from the wild. There is considerable evidence that delayed fertilization is the rule in some forms (e.g. Rhinolophidae) although second mating in the spring almost certainly occurs in some individuals.

Harrison (1958) records several cases of successive pregnancies (lactating females carrying a foetus) in the Nycteridae and Molossidae of Africa. Nevertheless the reproductive behaviour and physiology of bats are little known, and observations of breeding in captivity would be very valuable.

DISEASES AND PARASITES

Many bats die suddenly in captivity for no apparent reason and are usually dismissed as being too delicate. Several of the Emballonuridae and Natalidae as well the European lesser horseshoe bat *(Rhinolophus hipposideros)* come into this category.

Even if it can be persuaded to eat, the latter species seldom survives more than a few days in captivity, whereas the larger *Rm ferrumequinum* may be kept for long periods. The reasons for these failures may be quite simple but they are completely unknown at present.

Since most ' ats both feed and drink on the wing, they must show considerable plasticity of behaviour in order to live in cages. It is perhaps more surprising that most bats adapt so readily than that others are unable to survive.

The wings of captive bats often develop sores and swollen joints. These may be treated with alcoholic gentian violet, although if large areas are painted the bat may ingest too much when licking the wings, and may vomit. It is better to prevent the condition by diet, exercise, and control of humidity. Loss of hair is probably due to diet.

Ectoparasites

Bats harbour many different parasites although few of these are well known. The most obvious external parasites are fleas, ticks, and both Streblid and wingless Nycteribiid flies which run with dexterity through the fur.

They do not seem to cause much harm, but may be caught with forceps or killed by a light dusting with an insecticide. One animaldealer claims that Nycteribiids are essential to good fur-condition in *Pteropus,* but there is no evidence to support this idea.

Hazards to Man

Far more important are the human diseases known to be transmitted by bats. Chagas's disease, and possibly relapsing fever may be transmitted from infected bats by blood-sucking ectoparasites in tropical regions.

Histoplasmosis and dermatomycoses may be transmitted through the debris of bat caves and possibly through the urine or faeces ejected by flying bats. In the presence of very dense populations it is advisable to wear a light face-mask to prevent inhalation of such matter. A greater danger lies in being bitten by a bat infected with rabies. Several genera of neotropical frugivorous bats (including *Artibeus* and *Carollia),* as well as the vampire *Desniodus,* have been shown to carry forms of rabid myelitis, which can be transmitted to man through a bite.

At least one case of death from bat rabies (from a vespertilionid bite) has been confirmed in the U.S.A. and several suspected cases have been reported from Europe. Anyone intending to handle bats large enough to break the skin, or any bats from the neotropics, is strongly urged to wear stout leather gloves and to treat accidental bites with suspicion.

Special care should be taken with bats found to be sick on capture. Finally, histological examination of a large number of hats of many species has revealed no pathological infection of the ear. This is in striking contrast to the situation commonly found in many more familiar laboratory animals.

HIBERNATION

Bats of temperate latitudes may hibernate in captivity but the author does not recommend that this be enforced. Except for the study of the physiology of hibernation, a sleeping hat is not a very interesting subject.

No ill effects are detected if activity is maintained throughout the winter months and it is then easier to ensure that the animal's food and water requirements are properly met. If hibernation is desired the bat should be fattened by intensive feeding for several weeks.

Wintering conditions must simulate those normally selected, an even temperature a few degrees above freezing and 100-per-cent humidity, since the bat will not drink for several weeks and water-loss must be minimized.

Even then. the conditions cannot be very natural, and survival is not guaranteed. Wild bats would normally enter hibernation voluntarily and rather gradually, and several species occasionally emerge for flight even in mid-winter.

ANAESTHESIA AND EUTHANASIA

Intraperitoneal injection of one volume of pentobarbitone sodium (nembutal) in nine volumes of 10-per-cent ethyl alcohol induces a suitable state of anaesthesia in a dose of 30 to 50 mg per kg of body-weight. There is considerable individual variation in response, and the exact dose depends on the animal's general condition.

Body-temperature is of great importance. The bat must be wide awake and active before being injected and, when anaesthetized, must be kept at a constant body-temperature for reliable results.

Rectal temperature is conveniently measured by a thermocouple and should be kept at 37 to 40°C.

Overdosage with pentobarbitone sodium may be used to kill bats, and this may be administered to torpid animals if necessary.

10

The Fruitfly

Since the domestication of *Drosophila melanogaster* early in this century, and the nearly universal use of this species in genetic experiments, information on culture techniques and laboratory handling has accumulated rapidly.

Numerous species of *Drosophila* can now be maintained in culture on media more or less simulating their natural diets. Sturtevant (1937) summarized the historical development of such media from fermenting fruit in glass jars to the complex semi-synthetic types.

Laboratory procedures for handling the flies were being developed simultaneously and the items of equipment in current use are quite varied and often complex. Earlier versions of some basic apparatus were described by Bridges (1932) and in the sixth issue of DIS in 1936 which was devoted entirely to notes on culturing and handling *Drosophila.*

More recent descriptions of the methods employed are to be found in Spencer (1950), Strickberger (1962), and Demerec and Kaufmann (1962); these references should be used in conjunction with the present survey since they deal with some aspects of laboratory techniques that are not being discussed here.

CULTURE MEDIA

Drosophila larvae feed principally on yeasts; thus, most laboratory media are essentially yeast cultures on carbohydrate bases, stiffened with agar to facilitate handling. Media which support larval growth and development are satisfactory for the adults as well.

Species which frequent fungi in nature are often very difficult to maintain in culture, as are those which inhabit flowers. Other species, which do rather poorly on standard media, can often be successfully raised if relatively minor changes are made in the medium. Many such modifications have been described, mainly in the research and technical-notes sections of D1S.

Several species of *Scaptomyza, Chymomvza, Mycodrosophila,* and *Zaprionus* (family Drosophilidae) can be cultured on *Drosophila* medium, as can species of *Megaselia* (Phoridae). With more difficulty, a few species of *Milichiidae, Muscidae,* and *Svrphidae* can be raised, and one suspects that this list could be greatly increased if other groups of fruitgarbage-feeding *Diptera* were tried.

General Requirements

Optimal growth of *Drosophila* depends upon a suitable proportional relationship between the main classes of nutrients: carbohydrates, proteins, lipids, salts, vitamins, and water. Since these nutrients can be supplied in culture media by several different methods, there seem to be about as many different formulae as there are *Drosophila* laboratories.

The carbohydrate source of the medium may be in the form of sugars or of flours and meals or, most often, both. The sugar source may be one or a combination of sugars or syrups. The selection of a sugar source seems to be somewhat arbitrary.

Hassett (1948) found the order of usefulness of the common sugars to adult flies to be as follows: fructose > maltose > sucrose > glucose > galactose xylose > lactose. Larvae do not use maltose; otherwise the sequence is the same.

Sang (1956) found that fructose added after autoclaving of the medium gave better growth than when it was cooked with the medium. The optimal amount of fructose was given as 0·75 per cent (7·5 g/I) but larger flies were produced with 2·0 per cent.

The choice of a sugar source would seem to be much more critical in a medium which does not incorporate live yeast than for the yeast-culture type of medium in which the flies feed primarily on the yeasts rather than on other constituents of the culture medium. Table elsewhere in this chapter summarizes the food formulae of a number of laboratories and illustrates the wide variation in carbohydrate components.

The protein, lipid and vitamin source for *Drosophila* larvae is the live or dead yeast of the typical medium. In synthetic media these nutrients are usually supplied in the form of casein (or specific amino acids), cholesterol (or ergosterol), and vitamins of the B group.

Most media, after cooling in the culture containers, are seeded with a suspension of live yeast, either brewer's yeast, baker's yeast,

Table 10.1: Formulae of cornmeal media in current use.

	Water (ml)	*Agar (g)*	*Yeast dry (g)*	*Cornmeal (g)*	*Sugar*	*Propionic acid (m)*	*Other*
University of Wisconsin	1,000	16·0	30·0	82·5	111.0 ml molasses	4·4	
University of California (Davis)	1,000	15·8	20·0	79·0	105·0 ml molasses	6·8	
University of Texas—1	1,000	14·1	17·6	70·5	158·8 ml molasses	6·1	Ethanol (95 per cent), 5 ml
University of Texas—2	1,000	11·2	18·0	71·0	98-0 ml malt syrup	5·0	Soymeal, 11·3 g; dark Karo, 21·0 ml; Ethanol (95 per cent), 5 ml
Inst. Animal Genetics, Edinburgh	1,000	17·0	17·0	127.0	83·0 ml molasses	4.2	Nipagin—0.8 g
University of Kentucky	1,000	10·0	20·0	100·0	116.0 ml malt syrup	5·0	
Zool. Inst.,	1,000	12·0	10·0	135·0	70.0 g	5·0	(Nipagin, 1·0 g, when

(Table 10.1 contd.)

(Table 10.1 contd.)

	Water (ml)	*Agar (g)*	*Yeast dry (g)*	*Cornmeal (g)*	*Sugar*	*Propionic acid (m)*	*Other*
Zurich					sucrose		propionic acid is omitted)
Calif. Inst. of Technology	1,000	5·4	18·0	100·3	30·4 g sucrose	4·9	Dextrose, 60·4 g; phosphoric acid, 0·5 ml
Oak Ridge Lab., Tennessee	1,000	6·0	29·0	80·4	28·4 g sucrose	4·8	Glucose, 56·8 g; phosphoric acid, 0·5 ml; K, Na tartrate, 7·9 g; CaC 12, 0·49 g
University of No. Carolina	1,000	13·8	—	123·0	166·0 ml molasses	—	Oatmeal, 7·7 g; Tegosept, 0·9 g

isolated wild yeasts, or the Y-2 strain isolated by Wagner (1944). In general, non-fermentative yeasts are best. Since it is the yeast which serves as the principal food of the larvae, a good culture medium must also serve as an efficient yeast medium.

This is not usually a serious problem since there is a rather rapid natural selection of the yeast strains which will be successful on a given medium. Living yeast is not a necessity, however. Baumberger (1919) showed that autoclaved yeast was a completely adequate food for *Drosophila* larvae, and many laboratories now add dried, dead yeast to the medium either as a supplement to the nutrients derived from live yeast or as a substitute for the live form, Table elsewhere in this chapter.

Since the dry yeast does not mix readily with water, it is sometimes advantageous, particularly in large batches of food, to wet it by mixing with 95 per cent of ethanol before adding it to the rest of the mixture. The alcohol serves the additional function of aiding in mould and bacterial inhibition.

Sang, *et al.* (1949) suggested that the limiting factor for larval growth in a live-yeast medium was probably a general food-shortage which occurs on the third to the fifth day of development in a culture of *Drosophila melanogaster*.

Yeast supplementation at this time is advisable. Additional yeast suspension may be added directly to the food, but many workers prefer to soak sterile strips of paper in the suspension and then add the yeasted strips to the culture. Not only will the larvae crawl on to the paper to feed, but many will pupate there.

Larvae of most species do best in a moist, jelly-like substrate, and agar is most often used to produce the desired consistency. The amount varies from about 15 to 25 g per litre of water, depending upon various factors. Some workers, dealing with certain strains or species, prefer a softer medium obtained by using less agar.

Media having other thickening agents in them, such as cornmeal, cream-of-wheat, etc., require a little less agar. The amount of agar required tends to change with the relative humidity, so that during the most humid months as much as 4 g per litre more are required than during very dry periods.

The particle-size of the agar must also be considered. Finely powdered agar such as is routinely used in bacteriology is more easily put into solution than is agar of larger particle-size, such as flakes. By adjustment of the cooking-time, however, agar in any of its forms can be used satisfactorily. The pH of the medium, although not critical to the development of *Drosophila,* is important in other respects.

A favourable pH for yeast species is about 4·5 to 4·8. When live yeast is used, maintenance of such a pH enhances yeast-growth. Buffers may be used to maintain the pH, but they are probably unnecessary; Bridges and Darby (1933) showed that in typical culture of flies on a yeasted, unbuffered medium there is a fairly regular shift in pH.

Beginning with an initial value of about 5·8, it shifts to a minimum of 3·6 to 3·8 owing largely to accumulated acetic acid, and this is followed by a rise to from 3·8 to 4·8 as the more alkaline waste products of the flies accumulate.

The hydrogen-ion concentration is also important in determining the amount of propionic acid which is added to many food formulae as a mould inhibitor. The amount of propionic acid required decreases as the acidity increases.

For example, at pH 5·0 about 2 ml of the acid are needed per litre but 13 ml are needed at pH 6·0. The Y-2 yeast, isolated from cactus fruit by Wagner (1944), has been carefully strain-selected to grow optimally on propionic-acid media.

Subcultures of this strain and directions for its maintenance and use may be secured from the Genetics Laboratory, California Institute of Technology, Pasadena, California. Similar strainselection can be carried out when desired by isolating yeasts from the crops of wild-caught adult flies, and plating them out on samples of the preferred *Drosophila* medium.

Those strains which grow well on the medium should then be tested for nutritional completeness by careful feeding tests. For laboratory use, live yeasts are mixed in a thick suspension and then added to the food surface either with a glass dropper when only a few vials or bottles are involved, or by spraying trays of opened vials or bottles with a simple atomizer such as that used for perfumes.

The spray apparatus should be cleaned and sterilized after each use and stored in a clean uncontaminated place. Live yeast should be added only to well-cooled culture media and the food vials or bottles should then be stored for about a day at or near room temperature to permit the yeast to start growth before flies are added.

Food Formulae

There are three basic types of culture media currently in use; these are the banana-agar type, the cornmeal-agar type, and the synthetic or near-synthetic type.

The first two are regularly used for general maintenance of stock, for standard genetic experiments, and for population cage experiments. Synthetic types are intended primarily for nutritional studies and biochemical genetic experiments, since they are chemically defined or nearly so, and hence lend themselves to the maintenance of aseptic and axenic cultures.

Banana-agar medium

A banana-agar medium was one of the earliest agar-type media used for culturing *Drosophila.* It is still used in some laboratories, especially for species of the subgenus *Drosophila,* many of which are more difficult to raise on the more commonly used cornmeal medium.

In addition to bananas and agar, this medium contains dead yeast, a source of sugar, and a mould inhibitor. The surface of the cooled medium is seeded with live yeast. A successful formula for the banana-agar medium is given in Table elsewhere in this chapter.

Table 10.2: Banana-agar medium.

Water	1,000 ml	Sucrose	25 g
Agar	20 g	Propionic acid	4 ml
Banana (ripe, crushed)	300 g	Ethanol (95 per cent)	4 ml
Yeast (dried brewer's)	50 g		

To prepare, mix the *agar* with the water and bring to the boil. Continue heating until the agar has gone into solution. Add mashed bananas (about three medium-sized bananas weigh 300 g). Stir and cook for a few minutes.

Then add the yeast and sugar, wetting the yeast first with the ethanol. Stir and cook until the mixture is homogeneous and the sugar is completely dissolved. Allow to cool partially, then add propionic acid. Pour into vials or bottles and allow to cool gradually.

Cotton plugs should not be added at this time since water of condensation must have a chance to evaporate; a layer of thin sterile cloth may be draped over the vials to help avoid contamination. When the product is cool, spray it with live yeast-suspension, plug with cotton, and store for about 24 hours at room temperature before use.

The bananas should be very thoroughly mashed before being added to the mixture; this can best be accomplished with a Waring blender. Bananas can be kept frozen for long periods.

When they are allowed to thaw at room temperature, their consistency is such that very little mashing is needed. The freezing technique is especially useful in areas where a steady supply of bananas cannot be guaranteed.

The above formula makes about 11 of medium. If Nipagin (Tegosept) or Moldex is used as the mould-inhibitor, the addition should be made while the medium is being cooked (see below).

Cornmeal-Agar medium

The most widely used type of food medium is the cornmeal-agar medium. It is less expensive to make in large quantities than is the banana-agar medium and is quite successful in supporting *D. melanogaster* as well as other species of the subgenus *Sophophora (pseudoobscura, willistoni,* etc.).

Ten versions of the cornmeal medium currently in use are shown in Table elsewhere in this chapter. Although the relative proportions of the ingredients vary considerably, almost all the formulae contain the following basic constituents: dead, dried brewers' yeast, agar, cornmeal, a source of sugar, and a mould-inhibitor.

Some laboratories prefer another meal such as oatmeal or soya meal in combination with the cornmeal as a source of additional nutrients. A relatively simple cornmeal-agar formula for making about 1,000 ml of medium is given in Table elsewhere in this chapter.

Table 10.3: Cornmeal-Agar medium.

Water	1,000 ml	Dark corn syrup (Karo')	100 ml
Agar	15 g	Propionic acid	5 ml
Yeast (dead, dried brewers')	30 g	Ethanol (95 per cent)	5 ml
Cornmeal (maize meal)	100 g		

To prepare, mix the agar with about three-fourths of the water, boil, and continue heating until the agar has gone into solution. Add ethanol to dry yeast to wet it. Add the syrup and cornmeal to the yeast mixture, then gradually stir in the remaining water.

Add this mixture to the boiling agar-water solution. Boil gently for 10 to 15 minutes. Allow to cool partially and then add the propionic acid. Pour into culture-containers and allow to cool completely before spraying the surface with live yeast suspension.

If Tegosept (Nipagin) or Moldex is used instead of propionic acid as the mould-inhibitor, it should be added to the medium in alcoholic solution during cooking.

Other media

Through substitution of components and alteration of proportions, the possible number of variations of the banana-agar and the cornmeal media is nearly infinite. There are other types of media, however, which differ more radically from the standard types.

Several of these are described below, but no attempt is being made to list them all. An agarless medium was devised by Spassky and has been used at Columbia University, New York City, for many years.

It works well with many species of the subgenus Sophophora, e.g., melanogaster, pseudoobscura, and willistoni. The ingredients per litre of water are given in Table elsewhere in this chapter.

Table 10.4: Agar-less medium.

Water	1,000 ml	Molasses (unsulphured)	148 ml
Cream-of-wheat	133 g	Tegosept (10-per-cent	
Table salt	9 g	solution)	10.3 ml

Directions for preparing this medium are given by Strickberger (1962). A medium which can be prepared quickly in the field from all dry ingredients which have been previously mixed is especially useful.

In the laboratory the ingredients are weighed out and mixed in portions adequate for 100 ml of water, or for 250 ml, and so on. For field use it is necessary only to boil the required amount of water, stir in the ingredients, and pour the mixture into culture vials.

The only major difficulty relates to the mouldinhibitor, since Moldex and Tegosept are scarcely soluble in water and it may be inconvenient to carry along the alcohol needed to dissolve them or the liquid propionic acid which is also a commonly used inhibitor.

Crystalline sodium propionate appears to be an adequate substitute, but it does not seem to have been tried in this way. An example of a simple dry-ingredients medium (except for the mould-inhibitor) is that of David (1962), Table elsewhere in this chapter.

Table 10.5: Dry-ingredients medium.

Water	1,000 ml	Cornmeal (maize meal)	80 g
Agar	12.5 g	Tegosept (10-per-cent	
Yeast (dead, dried brewers')	80 g	solution)	5 ml

A synthetic culture medium was devised by Pearl (1926) to eliminate the need for bananas. It has since been altered in many ways, mainly by changes in proportions, the addition of other salts, and the addition of dried brewers' yeast.

One variation, currently in use at the University of Chicago and which might best be termed a '*semi-synthetic*' type, is given in Table elsewhere in this chapter.

Table 10.6: Banana-less medium.

Solution A		*Solution B*	
Water	800 ml	Water	200 ml
Agar	15 g	Potassium sodium tartrate	8 g
Yeast (dead, dried brewers')	15 g	Calcium chloride	0·5 g
		Sodium chloride	0·5 g
Sucrose	50 g	Manganous chloride	0·5 g
Potassium phosphate	1 g	Ferrous sulphate	0·5 g

Water is stirred slowly into the ingredients in Solution A. The mixture is then cooked by autoclaving for 30 minutes at 15 lb pressure. Then solution B is added and mixed. When the food has partly cooled, 5 ml of propionic acid are stirred in as the mould-inhibitor.

The medium is poured into containers and, when it is thoroughly cool, a live-yeast suspension is sprayed on to the food surface. During the winter months a further 100 to 150 ml more of water are added to Solution A as an adjustment to the climatic conditions of the Chicago area.

For precise nutritional studies and experiments in biochemical genetics, a chemically defined medium is required. However, since different species and different strains of a species have different nutritional needs, it is probably impossible to devise a synthetic medium which will permit normal growth of them all.

D. melanogaster has a normal larval-development time of 4·0 to 4·5 days on rich, well-yeasted media; in contrast, larvae on synthetic media often require twice as long, or still longer, to develop. One of the most efficient chemically defined media in this respect is formula C of Sang (1956) which could give completed development in 4·4 days at 25°C (77°F) with an optimal number (40 to 50) of larvae per culture.

Better known is the Hinton medium (Hinton *et al.*, 1951) which is now available commercially from Difco Laboratories, Detroit, Michigan. It is supplied in two forms, 3·34 g of agarless medium to be rehydrated in 100 ml of cold distilled water and then autoclaved to form a nutrient broth, and 4·84 g of an agar medium to be suspended in 100 ml of cold distilled water followed by autoclaving.

The final reaction of these media is pH 7·0. The medium can be altered to adapt it for other species, as was done by Jaegar and Jaegar (1957) for *D. willistoni*.

According to Sang (1956) the addition of whole casein to the Hinton formula increased its effectiveness; casein hydrolysate had no advantage over whole casein. For certain experiments, of course, this would destroy its usefulness in studying individual amino acids in the diet.

Food Modifications

In addition to the simple alteration of ingredient proportions, substitutions of sugars or meals, etc., culture media can be modified for special purposes. The addition of finely powdered charcoal or lampblack provides a useful dark background for making egg counts.

Since it is not easily kept in suspension while the medium is cooling, von Borstel and Fine (1961) recommend adding grape juice to provide a darkened background as well as additional nutrients. Colours as used for household food may also be used for this purpose, or for making different batches of media quickly recognizable.

The inability to raise many *Drosophila* species seems to lie in a failure of the females to oviposit. Oviposition can sometimes be stimulated by adding small amounts of the natural food, such as sterilized bits of mushrooms or of cactus fruits, to the surface of the medium. The seeding of poor cultures with wild yeasts, isolated either from the crops of the adults or from slime fluxes, serves the same purpose, as well as providing better nutrition.

The isolation of yeasts from other sources—fungi and flowers, for example—for use in *Drosophila* cultures seems not to have been tried. However, Lund (1954) has shown that decaying fungi (with an acid reaction) have a very high yeast content, and he also found numerous species of yeasts, although in much smaller numbers, in flowers.

His Table 39, based upon his own work, is quite impressive, listing 27 species of yeasts from flowers, 7 from agarics, 9 from tree exudates, 13 from insects, 5 from dung, 27 from soil samples, and 12 from fruits. There can be little doubt but that among these are the yeasts used by various *Drosophila* species in their wild environments.

Changing the food surface is also sometimes effective. Lightly scratching the food surface may result in large numbers of eggs being deposited in the scratched grooves on the medium. If the medium is allowed to harden on a slant and sterile sand or shredded paper towelling is added to the exposed portion of the bottom, larvae of some species seem to be more successful (G. L. Bush, personal communication).

The simple addition of sterile paper strips to the culture often results in a greater yield, especially with those species in which active larval movements tend to bury the puparia already present.

Food-Dispensers

Agar dissolves in water near the boiling-point, so that agar-type media require cooking at rather high temperatures to insure that proper jelling will take place. Cooking may be done in open pots or kettles, in pressure cookers, or in containers placed within pressure autoclaves.

Dispensing of the medium into vials or bottles is usually done by hand when small batches are involved, but some sort of a mechanical dispenser is preferred when larger amounts must be handled.

In many laboratories the hot medium is allowed to flow by gravity through a large-bore rubber tube, the flow being cut on and off by a hand-operated spring clamp or a stop-cock near the dispensing end.

Lewis (1954) described a pressure system using an army-type field autoclave of 25quart capacity. The autoclave was adapted so that compressed air could be used to force the hot medium through a tube connected to a controlling Morrison valve.

A similar but smaller installation was used by Vloeberghs (1960) in which an aquarium-sized air pump was attached to the dispensing container, with the output of medium regulated by a ground stop-cock.

If the contents of a culture-container are not poured out rather rapidly, the freshly cooked medium may cool enough to begin solidifying prematurely and may clog up the dispenser.

In the Vloeberghs apparatus mentioned above, spirally wound copper tubing with hot water flowing through it was twisted around the dispensing container to keep the contents hot. Traut and Mohr (1960) described another type of apparatus for keeping the medium hot during the bottling process.

The medium was poured into a large metal funnel which was surrounded by a 'ring heater for a water boiler'. Temperature was regulated by a rheostat with asbestos insulation. For rapid and uniform dispensing of large batches of medium, the pumping device described by Faberge and Cave (1952) is most efficient.

With a motor-driven piston-andcylinder system, squirts of medium are delivered at a regular rate. The amount of medium per squirt and the rate of delivery can be controlled so as to give, for example, 10 ml squirts at a rate of 100 per minute.

This device, which costs about $100 to construct, is in use in several laboratories. Blueprints for its construction are available from Dr A. C. Faberge, Genetics Foundation, University of Texas, Austin 12, Texas.

LABORATORY HOUSING

Environmental Requirements

The temperature at which cultures are maintained influences the length of life cycle, fertility, penetrance of mutations, repeatability of experiments, growth of yeasts and moulds, and rate of required stock-changing. For genetical studies of *Drosophila* 25°C (77°F) is generally considered to be the standard, a variation of ± 1°C (2°F) being acceptable.

Extremes of heat are much more damaging than are extremes of cold, especially when the adults are under the influence of ether. Young and Plough (1926) showed that *Drosophila* males can be rendered sterile by exposure to 31°C (88°F) for 10 days or less, though the females usually remain fertile. Such males will usually recover fertility after being returned to normal temperatures.

Generally speaking, higher temperatures favour more rapid maturation of cultures, while cooling tends to prolong development. Most healthy stocks can be kept for one or two months at a temperature of 18 to 20°C (64 to 68°F) without damage.

This is especially useful for reducing the frequency of stock-changing, for keeping duplicate stocks, for storing stocks during vacations, and for like purposes. Severe cold shocks will not harm most adults unless the exposure is unduly long. Novitski and Rush (1949) showed that males can survive for about 24 hours at 0°C (32°F) in laboratory tests, but etherized adults require about two hours to recover from the anaesthetic before being subjected to this temperature, or death will result.

The larger *Drosophila* laboratories install total air-conditioning with individual controls in each room. The ability to maintain different temperatures simultaneously is particularly advantageous since some mutant strains and species stocks from warmer climates do better at slightly higher temperatures, while others prefer a cooler environment.

The home window air-conditioning units are becoming increasingly popular for maintaining relatively constant coolness in isolated rooms; combination with a thermostatically controlled fantype electric heater permits year-round temperature-control.

For smaller installations, cultures can be maintained in thermostatically controlled chambers. Such a system can be installed at moderate expense when proper insulation and controls are available. When air-conditioning of any sort is not available during the extremes of summer heat, cooling can be accomplished by placing the cultures in pans of constantly flowing cool tapwater. Light influences the time of eclosion of adults, most emergence taking place within a few hours after sunrise, but for most species the amount of light available to laboratory cultures does not appear to be critical.

Some stocks have, in fact, been maintained for many generations in complete darkness. There are a few species, however, which wilt not mate in the dark. Cultures are most successful when the relative humidity within the vial or bottle is at or near 100 per cent.

Humidity within a culture vial is determined by a number of interrelated factors, such as the surface area of the food medium, water-content of the medium, volume of air within the container, ventilation, surrounding temperature, metabolic water contributed to or lost from the medium by the activities of the flies and larvae, and growth and fermentation of the yeast.

Soft foods (i.e., those with higher moisture and/or lower agar content) are generally the most favourable for larval growth, but they introduce difficulties in the manipulation of containers, the removal of adults, etc.

Although careful adjustment of the proportion of water in the food and careful choice of food-container size seem to be the best ways to maintain a satisfactory level of humidity, the addition of sterile filter-paper strips, absorbent facial tissues, cellucotton, paper towelling, or the like often helps to regulate relative humidity within a vial and in addition provides a convenient pupation site.

Culture Containers

The type of container used for *Drosophila* cultures is not critical, hence the factors of availability, cost, and the ease of filling and of cleaning are the first considerations. Most commonly used are glass shell vials of 8-dram (30-ml) capacity (25 *x* 95 mm) and half-pint (0·28-1) milk bottles.

For field use and for pair matings smaller containers, such as 3-dram shell vials, 5-dram plastic vials, or the creamers used in many restaurants, are better. The 5-dram plastic vials, being 70 per cent lighter than glass, are preferred for field use; for ventilation a 8-in (0·95-cm) hole is cut in the plastic lid (a standard cork-borer can be used) and a small cotton plug is inserted firmly into the hole.

Glass vials and bottles are routinely plugged with cotton-wool, which not only confines the flies but ensures adequate ventilation. The cotton-wool should be sterilized before use to avoid contaminating the culture with mould spores.

Many laboratories wrap the cottonwool plugs with a layer of thin inexpensive cloth to form permanent re-usable plugs. These plugs should always be autoclaved in a steam sterilizer before re-use.

Although it is seldom desirable, standard cardboard milk-bottle caps may be used. If the caps have been treated with wax some tiny holes must be punched for ventilation, or a larger hole plugged with cotton-wool must be added. Polystyrene caps can sometimes be found in a size which fits culture bottles.

The small tab on one side makes handling easy. A hole plugged with cotton-wool should be added for ventilation. The size and shape of the container limits the amount of culture medium which can be used and therefore the number of flies which can be supported.

The practical, functional food capacities of various containers are about as follows: 5 ml for 3-dram shell vials, 7 ml for 5-dram plastic vials, 10 ml for 8-dram (30-ml) shell vials and creamers, and 50 to 75 ml for half-pint (284-ml) milk bottles.

Containers of other sorts have been used for special purposes. Altenburg and Browning (1959) described a vat method of culture capable of producing up to 25,000 flies.

In a glass bowl 8 in (20 cm) in diameter containing a 4-in (19-mm) layer of well-yeasted food, as many as 10 broods of flies could be obtained if yeast supplementation was carried out at regular intervals. At the other extreme, inexpensive disposable paper containers can be used. Two types were described by Arnold (1958) and Moyer *et al.* (1961), and a third type is marketed by Wards Natural Science Establishment, Rochester, New York.

Discarded culture-containers should be sterilized by autoclaving as soon as this can be done; old containers left lying around serve as an unnecessary source of contaminating moulds, mites, and flies. Autoclaving also melts the medium, making thorough washing *of* the containers much easier.

The drains which carry off the waste should be flushed periodically with scalding hot water to prevent stoppage due to accumulated masses of agar.

POPULAR CAGES

The population-cage technique for maintaining artificial populations of different species of *Drosophila* was introduced by L'Heritier and Teissier (1933). Cages of nearly any size may be used, provided certain requirements are met. There must be some facility for adding culture media to the cage in measured amounts and at regular intervals. There must be adequate access for periodic removal of samples. There must be access for thorough cleaning between utilizations.

All parts must fit snugly enough to prevent accidental introduction of contaminating species. In addition some degree of ventilation should be provided for, as well as a reasonable degree of temperature-control. The following discussion illustrates three major types of such cages.

A modification of the L'Heritier-Teissier cage (Wright and Dobzhansky, 1946) has been widely used, and can maintain continuous populations of between 3,000 and 5,000 flies.

Various other modifications have been suggested; that shown in Figure elsewhere in this chapter is one of the late models capable of maintaining populations at the 10,000 level. The box measures 6 6 18 in (15 × 15 × 45 cm) and is made of 4-in (6·3-mm) clear acrylic plastic; at one end is a hole through which the parent flies are introduced and which is then closed with a rubber stopper.

At the other end is a ventilation opening fitted with 100-mesh screen. Food vials, inserted from below, are 9-dram (or 10-dram) plastic (Plastainer) vials obtainable from wholesale drug houses, and will hold up to 30 ml of medium.

The slightly thickened upper rim helps to hold the vials in place, but a tight fit is guaranteed by the insertion of a flexible neoprene gasket layer (shown in black) between two clear plastic layers, the vial holes being 1 in (2·5 cm) in diameter in the neoprene, and 1·5 in (3·8 cm) in the plastic.

The small figure to the right shows a hypothetical section through a vial hole and illustrates the contact of gasket to vial top. Before insertion of a fresh food vial, the food surface is scarified, several drops of live yeast suspension are added, and a sterile strip of paper (paper towel or filter paper) is dipped into a solution of mould-inhibitor (for example, 1 to 2 per cent of Moldex in 70 per cent alcohol) and pushed into the food mass.

The paper strip serves also as a favourite pupation site and absorbs any excess moisture which may develop. Several novel ideas on construction, incorporated in the above model, were described earlier by Wallace (1949) who designed a semicircular, plexiglass cage for exposing entire caged populations to repeated and constant doses of radiation.

The chief advantage of the L'Heritier-Teissier cage is the large size of the population which can be maintained. Smaller, less expensive, cages, however, are more useful for certain experimental designs. The cage described by Bennett (1956) is a good example, consisting of a polyethylene plastic box of the type used to store food in a home refrigerator.

Holes of an appropriate size are cut out on two opposite sides to permit insertion of standard glass culture vials containing the medium; ventilation holes are cut in the other two sides and are covered with cloth or fine mesh screen. The population size in a Bennett cage varies from about 300 to 2,000 according to Frydenberg (1962) who also shows a photograph of a cage in use.

A useful modification of the Bennett cage is described and figured

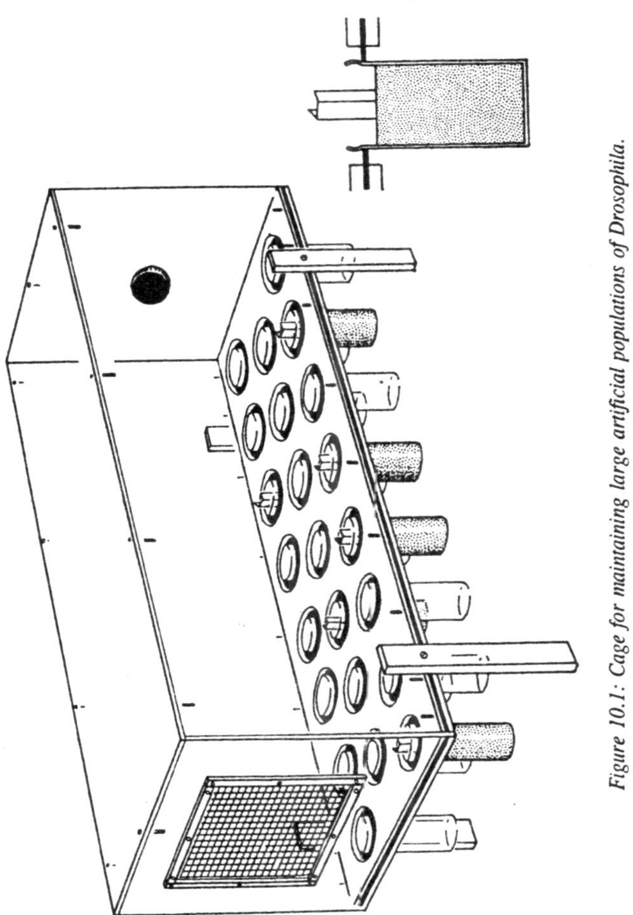

Figure 10.1: Cage for maintaining large artificial populations of Drosophila.

by Pipkin (1962). The culture medium is contained in four quart-sized (1.1-l) Mason jars attached horizontally to a central stainless-steel chamber, with an introduction hole and ventilation holes appropriately placed. Pipkin reports that soon after the second generation in such a cage the total number of flies reached a level of 2,000 to 3,000.

Simpler and smaller cages have been described by Reed and Reed (1948) and by Susman and Carson (1958), with consequent reduction in total population-sizes.

The Reed cage consists of two half-pint milk bottles connected mouth to mouth by a section of rubber automobile radiator tubing; the population is supplied with food by exchanging one of the bottles at regular intervals with a fresh one containing a suitable amount of culture medium. Ventilation is provided by a cotton-stoppered hole cut in the rubber tubing.

The Carson population chamber is a single standard Drosophila shell vial, 25 × 95 mm; by transferring the parent population (50 flies) to a fresh vial every two days and adding all offspring to the parent population, levels of 200 to 500 flies are obtained.

Sampling of larvae (for chromosomal analyses etc.) is simply done by removing one or more food cups. Removal of adult samples is more difficult. If the cage is covered to produce darkness and a bright light is placed near an appropriate exit hole at one end, the flies will move toward the light, especially if they are mechanically disturbed.

Simple aspirators of the type frequently used in entomology may be employed when small samples are needed. For larger samples a vacuum pump attached to a movable suction device is more practical. An example of such a device was described by Nicoletti (1957).

Observation Techniques

Most culture-containers are made from thick glass and do not permit of careful observations of the activities of the flies within. The thinner-walled shell vials are better, but optical irregularities in the glass often make them unsatisfactory; the same is true of most plastic containers.

A relatively shallow container, rectangular or round, temporarily covered with a thin plate of good glass, is quite useful. In addition to the inclusion of food samples etc. (depending upon the nature of the experiment), a piece of water-soaked absorbent cotton should be included to maintain the necessary high level of humidity.

A chamber of this sort, modified for the introduction of odour-laden jets of air, was described by Rizki (1957).

An ingenious technique for observing and recording the flight paths of insects was described by Kellogg and Wright (1962); its use with *Drosophila* to study their olfactory guidance was discussed by Kellogg *et al.* (1962).

For the observation of larvae, shallow food-containers are best. Lindsay (1958), for example, described the use of petri dishes to study the yeast preferences of *D. pseudoobscura* and *D. persimilis* larvae. This technique could easily be adapted for other types of experiments.

SPECIAL LABORATORY TECHNIQUES

Anaesthetization

Adult flies may be shaken from one food container to another by hand, or their positive phototropism may be utilized by allowing them to move from one container to another toward a bright light. However, for sorting out the species from wild collections, for separating sexes, etc., the flies should be anaesthetized.

Ether is most commonly used and many kinds of etherizers have been invented. A number of them are shown by Demerec and Kaufmann (1962). The etherizers shown in Figure elsewhere in this chapter illustrate two simple, basic types; the one on the right consists of a glass jar containing a wad of cotton or gauze which is soaked with ether.

The fly-containing chamber, punctured with numerous small holes, is attached to a fitted lid and suspended in the ether-air mixture of the jar. The upper portion of the fly chamber, often mechanically shaped of plastic, rubber, or metal, is funnel-shaped so that the various sizes of culture tubes or bottles, when inverted to shake the flies into the etherizer, will fit as snugly as possible inside this portion of the chamber.

The lower section of the fly chamber (i.e., that part which actually contains the flies) may or may not be made of the same material as the upper funnel-shaped part; glass, plastic, and metal are the materials most frequently employed.

Transparent substances are usually preferred since they permit one to observe the flies in order to judge the degree of etherization and determine the ideal moment to pour out etherized flies.

The etherizer shown on the left of Figure elsewhere in this chapter allows greater precision in adjusting the amount of ether administered; one or more squeezes of the rubber bulb will force relatively uniform increments of ether-air mixture through the connecting rubber tubing into the vial containing the flies.

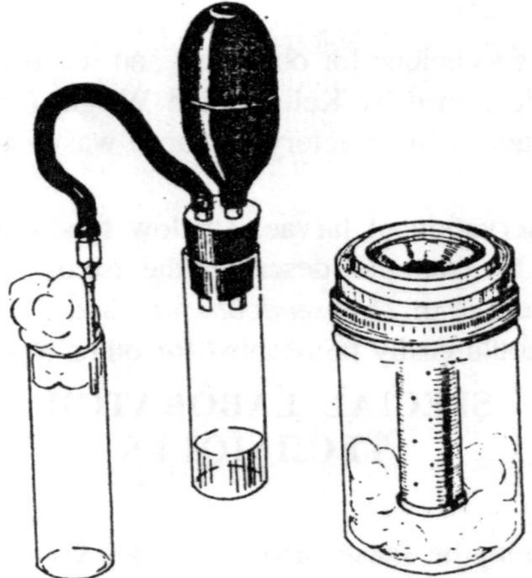

Figure 10.2: Two basic types of standard anaesthetizing containers (etherizers).

The tube ends with a large-bore injection needle. By using varying amounts of ether in the container, the relative proportion of ether and air is easily adjusted.

In etherizing flies certain factors must be kept in mind. Different species of *Drosophila,* or even different strains of a species, tolerate different dosages of ether; one must learn from experience how much ether a given strain can take without being killed.

Etherization at unusually high or low temperatures is more often fatal. This is one reason why air-conditioned laboratory rooms are desirable. Newly emerged adults are less resistant to ether than are older flies.

A common source of trouble is the heat generated by a strong microscope light. This is easily remedied by passing the light through a water-filled round flask so adjusted in height that the cooled beam falls on the etherized flies.

From among a large number of etherized flies some will, undoubtedly, begin to recover too soon and by their movements interfere with the work. To quiet such flies several sorts of re-etherizers have been devised.

Two of these are figured by Demerec and Kaufmann *(1962)*. The simplest is a glass funnel with absorbent cotton or cloth stuffed into the narrow neck; a few drops of ether are placed on this cotton plug and the funnel is inverted over the flies.

Additional drops of ether can be added as needed. A durable

polyethylene etherizer, which can also be used as a re-etherizer, is now marketed by General Biological Supply House, Chicago, Illinois.

Etherization is employed by most workers to separate the sexes soon after emergence so that virgin flies will be available for experimental matings. A method of virginizing mated females was described by Novitski and Rush (1949); they reported that *D. melanogaster* females were de-seminated by exposure to subzero temperatures.

An effective exposure programme was - 10°C (14°F) for 10 minutes. Males so treated recovered partial fertility after several days. When ether is undesirable, carbon dioxide may be used as an anaesthetic. A slow steady stream of the gas from a pressure tank is allowed to pour into an open plate (e.g., half a petri dish) containing the flies; they will remain motionless as long as the gas surrounds them, but will recover quickly when removed from it.

To immobilize individual larvae, as for *in vivo* experiments, the method suggested by Meyer (1957) is recommended. Larvae which have been placed in a drop of water are put into a container filled with nitrogen gas.

After about 5 minutes the larvae are limp and stretched out at full length. They appear to recover without any ill effects when brought back into air. Flies to be discarded after etherization and study should be dumped into a morgue.

This usually consists of a bowl or can, partially filled with a light-weight oil or drained engine oil. If a narrow-mouthed container is used, it may be best to have a receiving funnel attached to the opening.

Mites and Mite-Control

Several mite species live successfully on *Drosophila* media, but only one has been proved to be a deleterious pest. The general nature of mite infestations, their life cycles, optimal culture conditions, etc. have been discussed by Spencer (1950), Hughes (1946), and Binder (1957).

The most serious pest is the parthenogenetic species, *Histiostoma laboratorium* (sometimes called *H. genetica);* it should be eliminated from an infected laboratory as quickly as possible. The most important factor in mite-control is general cleanliness, including thorough sterilization of materials and equipment and prompt disposal of discarded cultures.

In addition two miticides, benzyl benzoate and Aramite, are currently in use. Benzyl benzoate is commonly used as a 5 to 10-per-cent solution in 95-per-cent ethanol. The cloth wrappers of cotton plugs from culture vials are soaked in the solution and allowed to dry; the deposit remaining is adequate to kill mites which crawl across the cloth.

This solution may also be used to wipe table tops, storage racks, and other equipment. The residual effects of the benzyl benzoate solution make it more effective than a weak (1:100) solution of phenol or lysol for scrubbing equipment.

Effective mite-control using the organic miticide, Aramite (2(p-tert-butylphenoxy) isopropyl-2-chloroethyl sulphite) was described by Stone and Zimmering (1951). Adult flies from infected cultures are transferred to a fresh dry vial along with about 150 mg of the powder (Aramite-15W has 15 per cent active ingredient in a wettable powder carrier); the vial is then shaken so that each fly becomes covered with the powder.

Cotton plugs also may be dusted with the miticide. These authors reported no ill effects on the species used, *D. melanogaster;* however, before using this product on other species it would be wise to make a trial run. Free moisture should be avoided since wetted Aramite will kill *Drosophila.*

Small local treatments will not eradicate mites when an entire laboratory is infected. Such rooms should be fumigated whenever this is practicable. One reasonably safe method, described by Stone and Zimmering (1951), involves a mixture of 3 parts of ethylene dichloride to 1 part of carbon tetrachloride, about 15 lb (6·8 kg), of the mixture per 1,000 ft^3 (28 m^3) being required.

The miticide, Aramite-15W, may also be used as a spray for entire rooms. The wettable product is added to water at the rate of 1 lb (454 g) per 25 gal (115 l) of water; I gal (4·5 l) of the diluted mixture will treat up to 500 ft^2 (46 m^2) of wall surface, depending upon the absorptive capacity of the surface. The safety precautions listed on the container should be observed. Aramite is sold by the Naugatuck Chemical Division of the U.S. Rubber Company, Naugatuck, Connecticut.

Moulds and Bacteria

Moulds are not often a problem in rapidly growing *Drosophila* cultures, but in poorer cultures, weak mutant stocks for example, the growth of moulds can be a serious problem. Control is accomplished in most laboratories by the use of a chemical inhibitor. Only the inhibitors most commonly used will be discussed here.

Two derivatives of p-hydroxy benzoic acid may be added to the culture medium to inhibit mould growth. These are the methyl ester, going by the trade-names Tegosept-M, Nipagin, or Methyl Parasept, and the sodium salt, called Moldex.

These are only slightly water-soluble; hence the usual practice is to prepare a 10-per-cent solution in 95-per-cent ethanol, then add measured amounts of this solution to the food medium during the cooking process. The amount required is 7 to 10 ml of the solution per litre of water used in the medium (i.e., 0·7 to 1·0 per cent).

At higher concentrations, e.g. 1·5 per cent, these substances are believed to retard the growth of both flies and yeasts. Propionic acid is widely used as the mould-inhibitor, especially in the United States. Most laboratories use about 5 ml of the acid, technical grade, per litre of water used in foodmaking.

As was stated earlier, however, the amount required depends upon the pH of the medium, a lower pH requiring less propionic acid. Mittler (1947) discussed the reasons for this variation and provided a formula for determining the amount of acid needed for a given pH.

He found that at pH 5·0, 2·1 ml of acid per litre of food would inhibit moulds, but at pH 6·0 inhibition required 13·2 ml. Lewis (1960) has recommended lowering the pH by the addition of phosphoric acid, thus reducing the amount of propionic acid which must be added.

The food medium should always be allowed to cool slightly before the propionic acid is added. Sodium propionate should be an equally effective mould-inhibitor, and has been used by Lund (1954) at a level of 0·25 per cent (2·5 g/l) to control moulds and bacteria in cultures of wild yeasts.

It is highly soluble in water, is most active at an acid pH, and is considerably more convenient to handle than the liquid acid. Although serious bacterial infections are greatly reduced by the use of mould-inhibitors, bacteria are occasionally introduced into cultures by the flies, especially those recently captured in the wild.

Excessive liquefaction of an agar-type medium may be due to such a bacterial infection, although there are other conditions which have the same effect. Some of these are: excessively low pH, too rapid cooling of freshly-cooked medium, re-cooking of jelled medium, improper water-agar proportions.

With other types of contaminating bacteria the food may develop a sticky or slimy surface and turn dark. Unless treated promptly (or discarded) such cultures are apt to be lost. To deal with this problem Spencer (1950) recommended frequent transfers of adult flies on to fresh food and the addition of large amounts of live yeast to larval cultures.

A potent solution of aureomycin and streptomycin (non-medical grades are satisfactory) added to the food surface gives more rapid control. A solution found to be satisfactory at the University of Texas laboratory is the following: about 350 to 375 mg of aureomycin are mixed with about 700 to 750 mg of streptomycin in 50 ml of distilled water.

Absolute accuracy of measurements is not critical, although a ratio of 1:2 of aureomycin to streptomycin is desirable. Under refrigeration, this solution retains its efficiency for at least a month.

A medicine dropper is usually employed to add the antibiotic solution to the cooled food-surface, one drop per culture vial and 2 to 3 drops per bottle being sufficient. An atomizer could also serve as an efficient dispenser. Several such treatments are usually necessary to rid a culture of bacteria.

Other Pathogenic Organisms

Although moulds, bacteria, and mites present the biggest problem in *Drosophila culture*, other pests are sometimes encountered, especially when wild-caught flies are brought into the laboratory. Nematodes are occasionally introduced with flies inhabiting slime flux or fungus, and in time form dense, tangled masses on the food-surface.

There is no known remedy other than to discard the infected vials or cultures. Several sporozoan and fungus parasites have been encountered, living either in intestinal cells or within other internal organs. For example, Chatton (1913) described a parasitic yeast, *Coccidiascus legeri* (Endomycetaceae), living in the cells of the intestine of *D. funebris*.

This yeast produces 8 needle-sharp spores in each banana-shaped ascus, and in this respect somewhat resembles a sporozoan. The most serious sporozoan parasite, causing male sterility, was discussed by Wolfson *et al.* (1957) who made, however, no suggestions for a remedy. Intestinal flagellates also have been observed in *Drosophila,* but there is no evidence of damage to the flies.

Several species of hymenopterous parasites live on *Drosophila,* but they are seldom a problem unless one brings wild pupae into the laboratory. These have been discussed in detail by Spencer (1950) and by Young and Meyer (1954), the latter reporting an infection with *Pachvcrepoideus dubius* (Pteromalidae) which reached a level of 80 per cent of the pupae in a culture.

11

THE HOUSEFLY AND THE BLOWFLY

THE HOUSEFLY

With a Geographical range extending from the sub-polar regions to the tropics *Musca domestica* is probably the most widely distributed insect in the world. It is also perhaps the most important insect available to the laboratory worker for the routine testing of insecticides, insect repellants and attractants.

There is no doubt that it has come to be regarded as indispensable for the standardization of contact sprays (West, 1951). Houseflies are also employed extensively in zoos and laboratories as a live food for certain carnivorous insects, reptiles and birds.

The culture of *M. domestica* in the laboratory is simple. The flies breed rapidly and are extremely prolific. Disease is no problem and healthy stocks can be built up and maintained over long periods without difficulty. Several media for rearing flies have been used, ranging from pasteurized horse manure and yeast to quite complicated formulae.

It is, however, important to emphasize the point made by Parkin and Green (1957) that once a medium and methods of rearing and handling have been found satisfactory, they should be adhered to, or excessive variations in the physiological behaviour of the adults towards insecticides may complicate tests.

Obtaining Stocks

Stocks to start the cultures may be obtained either by acquiring pupae or adults from a laboratory already engaged in their production, and this is the safer method, or by setting traps, at least during the warmer months of the year, to catch specimens around farms, poultry houses and stables.

If the flies are wild-caught, they should be from an area where residual insecticides have not been used, to avoid introducing a strain of flies having some resistance to certain insecticidal compounds. Descriptions and keys for identifying the wild-caught flies are given by Hewitt (1914), Patton (1930), Herms (1950), and West (1951).

Housing the Adults

If flies are to be produced on a large scale and for insecticide-testing, the culture room should be sufficiently spacious to accommodate the whole of the equipment and have a controlled temperature of 27·5°C ± 1°C (81°F ± 2°F).

The relative humidity should be maintained at between 50 and 55 per cent, but should not vary more than 3 per cent. The ceiling and the walls of the room should be free from cracks and crevices and preferably sealed with several coats of hard gloss paint or enamel to facilitate regular cleaning with detergent and warm water.

White or cream paint has the advantage of showing up dirt or unwelcome insects or parasites which may find their way into the room. The floor should be concrete, preferably coated with either asphalt or bitumastic, thus avoiding any holes or cracks in which spilt food could lodge.

Sunlight or even daylight is not really necessary, in fact windows are to be avoided as they give rise to unnecessary heatloss and become damp with condensation during cold weather.

Most workers find that fluorescent lighting is to be preferred because of the low output of heat. Parkin and Green (1957) used an automatic time switch, giving nine hours' light each day.

It is important to avoid any tendency for the atmosphere in the room to vary in temperature and humidity at different levels, and this can be overcome by the use of a small fan to keep the air continually circulating. For the same reason racking should be used rather than solid shelving.

Mating and Oviposition

The simplest and cheapest type of cage for mating and oviposition is made of wood, and for convenience is not more than 28 dm^3 in volume. Wire or terylene gauze should be used for the sides and top, and the bottom should be of plywood.

One or two sleeved holes should be fitted into the gauze to allow the flies to be put in and food to be changed. Cages of this size will accommodate about 600 flies. These cages are cheap and easy to construct and are extremely satisfactory except that they are difficult to keep clean for long periods without deterioration, and so have to be periodically replaced.

Basden (1947) described a more robust type of wooden cage which can be dismantled for cleaning. The measurements are approximately 45 × 45 × 45 cm and it consists of two side frames with bolts protruding from the front, back and top.

The other three corresponding frames are fixed on to these bolts with the aid of wing nuts. The plywood floor fixes into retaining grooves at the base of the frames. Brass wire gauze (6 meshes per cm) or perforated zinc is used to fill in the side, top and rear frames. The front consists of an upper portion made of clear glass, while the lower half consists of plywood into which two holes, approximately 13 cm in diameter, are cut.

These holes are each fitted with a metal flange 13 cm in diameter and 2·5 cm in height on to which muslin sleeves 30 cm long are fixed by means of elastic bands.

Although these are excellent cages, they are rather heavy to handle and the tendency of the wood to warp sometimes makes them difficult to assemble. Parkin and Green (1957) therefore used Basden's original principles and designed an improved cage of 18-gauge aluminium.

The sides, front and back are cut and bent from a single sheet, the overlapping 1·25 cm of the two free edges being riveted; the jointed corners of the cage are then filled with solder to give a smooth internal finish.

The solid metal bottom and the top frame are cut 1·25 cm oversize on all edges. These edges are then bent at right angles and riveted to the upright sides of the cage. A sliding glass door forms the upper half of the cage front, while the lower half of the front is similar to Basden's, being finished off with two 13-cm flanged holes to which muslin or terylene sleeves are attached.

The same material is used to fill in the top, back and side frames where it is held in place by angled strips and springs.

Care and Feeding of the Adults

The number of flies needed to stock the oviposition cages is not critical but depends largely on the number of eggs required. A minimum of 50 cm^3 of space per individual is a good guide. The easiest way to stock the cage with a given number of flies is to insert a shallow open-topped container into which clean pupae have been counted or measured.

Immediately the flies start emerging, fresh food and water must be given and kept constantly available. The water can be given in the form of a fountain. Peterson (1959) describes various types but the one most commonly used is an upturned beaker of water resting on a shallow pad of cotton-wool which is pressed into a petri dish.

The flies can rest upon the damp cotton-wool between the edge of the petri dish and the beaker, and will not drown. Alternatively ajar of water can be used, through the lid of which a thick absorbent cotton wick protrudes.

Sugar should be provided as a thin layer in a shallow dish. Cotton-wool pads soaked in a mixture of equal parts of milk and water should be placed inside the cage for oviposition. These should be changed daily as they quickly become sour in the high temperature.

Souring can be delayed by adding 1 part of 40-per-cent formaldehyde to 40 parts of the diluted milk. The first eggs will be laid on the cotton-wool between 4 and 5 days after the flies emerge and egg-laying should continue for a period of up to 10 days.

Treatment and Transfer of Eggs

The eggs laid on the milk pads must now be transferred to the rearing medium. To estimate the number of eggs available, they may be detached by gently shaking the cottonwool pad under water. Decant the supernatant liquid and replace with clean warm water.

Transfer the eggs in suspension to a calibrated tube, allow to settle and the number present can be closely calculated. A volume of 0·1 ml contains approximately 700 eggs. Care should be taken not to stir or shake the suspended eggs unnecessarily as this may result in high mortality.

If there is no need to estimate the number of eggs, the soaked pads with their attached eggs are removed daily from the oviposition cages and transferred to jars of the rearing medium.

The jars are then closed with muslin tops secured with elastic bands; the depth of medium in the jars should be about 10 cm. Glass pickle jars are used and they allow the progress of the larvae to be followed carefully.

Rearing-media for the Larvae

Various media have been used since 1927, when Glaser discovered that continuous rearing of houseflies throughout the year was possible with a mixture of horse manure and bakers' yeast. Richardson (1932) published details of a synthetic medium consisting of wheat bran, alfalfa meal, bakers' yeast, malt extract and water which is more pleasant to handle and less liable to infestation by mites.

The official method of the National Association of Insecticide and Disinfectant Manufacturers for rearing houseflies for Peet Grady tests (Anon, 1951) is substantially as follows: to a mixture of 1,200 g of soft wheat bran and 600 g of alfalfa meal add a suspension containing 30 g of bakers' yeast and 50 ml of malt extract in 2,700 ml of water and mix thoroughly to give a loose texture.

The prepared medium is placed loosely in cylindrical battery jars 15 cm in diameter and 23 cm high. Each jar is seeded with about 2,000 eggs and covered with a single thickness of cheesecloth. In Great Britain grass meal has successfully been used as a substitute for alfalfa (lucerne) meal.

A formula recommended by many laboratories consists of 3 parts by weight of dried yeast powder, 4 parts dried malt extract, 60 parts grass meal, 120 parts toppings, mixed with 390 parts by weight of water.

The medium which has proved successful at the London Zoo for some considerable time, and is still in use, consists of a mixture (in parts by volume) of 2 parts fish meal, 2 parts ground rabbit pellets, 2 parts dried yeast, 1 part sugar or molasses, 6 parts bran. Mix with enough water to make the whole damp but not wet.

The jars should all be clearly marked with the date when the eggs were laid, the number of the oviposition cage, the strain of flies and the date when the pupae and flies are expected to be available.

Pupation, and Care of Pupae

About 8 days after the eggs are laid, the larvae migrate to the upper layer of medium and pupate. If the correct quantity of water was used in preparing the medium, the top centimetre or so in each jar should by then be dry.

One or two days later, the upper portion of the medium containing the pupae is removed. Either the flies can then be allowed to emerge directly from the medium into an emergence cage, or the medium can be allowed to dry thoroughly before being broken up so that the pupae can be separated by either sieving, rolling on a sloping board, or winnowing with the aid of an electric fan.

Alternatively the contents of the rearing jars can be emptied gently into a basin of warm water (25 to 28°C, 77 to 82°F) nine days after the eggs were laid. If the whole is allowed to soak for a few minutes, most of the crusty portions of the medium can then be easily broken up by hand.

This frees the pupae, most of which will float at the surface. They can then be easily collected with the aid of a small scoop made from wire gauze or perforated zinc. They should then be washed and spread on absorbent paper to dry off. When floated off, the pupae will mostly be reddish-brown but will darken as they dry.

At this stage the pupae are in a very sensitive condition and must be handled with great care to avoid damage and the production of abnormal adults.

Transferring the Adults

It is possible to avoid much of the handling of adult flies when stocking oviposition and holding cages, by counting pupae and placing these in the cages to emerge where they are required. Parkin and Green (1957) describe a useful holding cage for approximately 100 to 150 flies.

They use circular-knitted mutton cloth (cotton stockinette) which can be bought in rolls. A piece of this is drawn over a cylindrical 23 × 15-cm wire frame and then knotted at the top and bottom to retain the flies. The necessary food, i.e. milk diluted with 50-per-cent water, is given in a glass tube plugged with cotton-wool soaked in the mixture.

When the flies have to be examined, counted or otherwise handled and this cannot be easily accomplished, their lively movements can be reduced by chilling. It is possible to use certain anaesthetics but this requires careful judgement to avoid losses, and recovery time is usually lengthy.

Anaesthetization can, however, be easily achieved and controlled with the use of carbon dioxide. Recovery in fresh air is fairly quick. Although all precautions are taken to avoid escapes into the fly room, it is inevitable that the odd one or two find their way out.

These can be trapped in the balloon type of trap made of wire mesh by using a small piece of meat or fish as bait, or by providing a saucer of milk in which has been mixed 2 per cent by volume of 40-per-cent formalin, or by a syrup containing 1 per cent of malathion applied to glass plates or cotton gauze.

Parasites

It is important to prevent parasites or predators from entering the breeding room, and so it is best to avoid using wild-caught pupae to start a culture. There is always a possibility that wild pupae will come already infected with parasitic Hymenoptera.

Wild-caught flies also often carry mites which are sometimes very difficult to eradicate from the breeding room. Newly caught flies should therefore be quarantined and the eggs they produce should be used only after careful washing and examination under the microscope.

Certain species of fungi are parasitic on Diptera, but the only form encountered commonly is *Empusa muscae,* which occurs both north and south of the equator, and frequently assumes epidemic proportions among wild flies during the autumn.

Most of the hymenopteran parasites that attack *Musca* and its relatives belong to the Cynipidae or the Chalcidoidea. The former mostly attack the larval stages, whereas the Chalcids are pupal parasites.

In Great Britain much trouble has been caused by the small Chalcid *Mormoniella vitripennis* in laboratories producing *Musca,* and also commercial bait farms breeding *Calliphora* and *Lucilia.*

Fortunately it cannot locate pupae covered by I mm of sand (West, 1951), and attacks have been overcome at the London Zoo by this means together with the use of a substitute breeding-room for a period of six weeks; meanwhile the original breeding-room must be thoroughly cleaned and washed out with hot water and detergent.

THE BLOWFLY

Blowflies of the genera *Calliphora, Lucilia,* and *Phormia* are extensively bred and employed in toxicological and physiological investigations, although perhaps not quite so commonly as the housefly.

Possibly the biggest demand for blowfly larvae is as bait for anglers or as live food for birds, fish, and reptiles. Like the housefly's, their short life cycle and the fact that they can be bred all the year round makes them ideal for laboratory work.

Obtaining Stock

Stock can be obtained either from a laboratory engaged in culturing blowflies, from a fish-bait farm, or from nature. Adults can be caught with a balloon-type trap exposed near a slaughterhouse or boneyard, and eggs can be obtained by exposing pieces of liver, beef or fish head for oviposition by wild flies.

Whichever method is used, a number of different species will probably be obtained and identification will be necessary. The keys and descriptions published by Seguy (1923 *et seq.*), Richards (1926), Hall (1948) and Roback (1951) will be found useful.

Housing the Adults

Accommodation similar to that used for the housefly is required, but it is important that all inlets and outlets to the room should be screened to avoid contamination by gravid females of *Calliphora erythrocephala* from outside.

Owing to the smell of the larval food all air should be discharged outside the building, or a large adequately ventilated cabinet should be used for rearing. The room should be maintained at a temperature of 25 to 27°C (77 to 81"F) with a relative humidity of 50 to 60 per cent.

Mating and Oviposition

It is possible to breed certain species of blowfly under conditions which avoid the smell and other undesirable features associated with putrifying food. Elaborate methods have been developed by White (1937) and by Haub and Miller (1932) for the aseptic culture of *Phormia regina* larvae.

Hill, Bell and Chadwick (1947) describe a method based upon the work of Brown (1938) but with extensive modifications for the breeding of *Phormia regina.* The medium consists of powdered casein, brewers' yeast powder and powdered agar in the proportions of 30:3:1, to which is added lanolin and a solution containing phosphate.

The adult females are allowed to deposit their eggs on ground horse meat, from which they are removed with forceps or needles; when handling eggs it is advisable to use moistened instruments. At 25°C and relative humidity 70 per cent the eggs will hatch in less than 24 hours.

The larvae will cease feeding at about the eighth day and pass into a prepupal or resting stage, after which they will soon pupate. The pupal stage lasts five to six days and the flies emerge on the fifteenth or sixteenth day after the eggs were laid.

It is, however, only fair to say that, although this method is quite successful, the flies breed more prolifically on a meat or fish diet. Breeding-cages are similar to those used for houseflies. The number of eggs required will determine the number of adults used to stock a cage, but it is suggested that these should be not less than 200 nor more than about 800.

A pan of sugar must always be available in addition to a water fountain of the type used for houseflies, or a water bottle provided with an extruding cotton wick. A piece of moist ox liver or ground lean beef weighing about 40 g is placed in the cage daily.

The surface of the liver can be scored with a knife to provide moist furrows in which the eggs will be laid. In *Lucilia* and *Calliphora* the first eggs are laid four to five days after the flies emerge from the pupae, in *Phormia* after seven days.

Treatment and Transfer of Eggs

The eggs are attached to each other in batches and will not usually separate in water, but if necessary they can be freed by immersion for not longer than 10 minutes in a (-percent solution of sodium hydroxide. A suspension of eggs in water can be measured in a culture tube.

A volumeof 0·1 ml contains approximately 650 eggs of *Lucilia sericata* or 350 eggs of *Calliphora ervthrocephala*. The eggs are usually left on the liver or meat to hatch, the pieces being cut up into portions, each carrying about 600 to 700 eggs; these are kept in closed dishes until the eggs hatch.

Each piece is then added to a culture jar containing more food. The eggs should be covered in order to maintain the humidity essential for hatching. The time taken for the eggs to hatch varies according to species. At 24°C (75°F) *C. errthrocephala* takes 3 to 8 hours and *L. sericata* 10 to 14 hours, whereas *Phormia terrae-novae* takes as long as 19 to 25 hours. *Calliphora* sometimes deposits first-stage larvae.

Rearing-media for the Larvae

Various high-protein foods can be used for larval cultures. Frings (1948) used fermenting dog biscuits, and Dorman, Hale and Hoskins (1938) used fish heads to breed *Callitroga, Lucilia, Sarcophaga* and *Phorinia*. Minnich (1937) used fish heads lying in soup plates.

Three such plates were stacked one above the other on a wooden rack which rested on a layer of soil inside a rearing-can ; the latter was fitted at the top with a manually operated damper to regulate the ventilation and humidity.

Pupation took place in the soil at the bottom of the can. Ox liver or ground horse meat is an excellent food for the larvae. The resulting smell can be kept at a relatively low level by feeding fresh or chilled food, and not giving more food than the larvae will consume. Several small cultures are more easily managed than a large one.

Glass pickle jars can be used for culturing, the small pieces of liver or meat carrying the eggs being placed on a larger piece (about 300 g in weight) which rests in the middle of the base of the culture jar.

A small quantity of sawdust, dry peat, or vermiculite should then be sprinkled around the food to absorb any exuding liquid. More food can be added if necessary.

Pupation and Care of Pupae

In most species the larvae migrate from the food when fully grown. Just before this stage they should be provided with sufficient material in which to pupate.

Here again a further quantity of sawdust, peat or vermiculite should be added to a depth of about 3 cm. *Calliphora* and *Lucilia* larvae feed for four to five days, *Phormia* for six to seven days. Pupae can be separated by the methods described for the housefly.

Transferring the Adults

Here, as with houseflies, it is possible to avoid undue handling of adult blowflies, by counting out pupae and placing these in the selected cages to emerge where they are required. If it is really necessary to handle the adults, and they are very lively, they may be quietened down by chilling or by the use of carbon dioxide.

12

The Rabbit

The European rabbit has been domesticated for many hundreds of years, and has been exported in a domesticated form to countries in most other parts of the world in the wild the species is a highly successful one, and has been found able to thrive over a considerable latitude in both hemispheres.

During the present century, in particular, many breeds and varieties have been developed, and at the present time in Great Britain there are some 35 breeds with more than double that number of colour varieties. The *Official Guide Book* of the American Rabbit Breeders Association lists 66 breeds and varieties of rabbit.

Variation in size (2 to 15 lb, 1 to 7 kg, at adult weight), in conformation (extreme raciness to extreme cobbiness); in ear-length (2 to 13 in, 5 to 32 cm) and in colour allow a wide choice of different varieties for different purposes.

Certain general characteristics make the rabbit peculiarly well suited to certain laboratory proceedings. As the doe ovulates only after mating, the time of ovulation can be determined with great accuracy, and dated embryological material can easily be obtained.

The large ear veins, which facilitate the withdrawal of blood, make the rabbit one of the most satisfactory subjects for serological work. It is well suited also to teaching purposes and its use for the diagnosis of pregnancy by the Friedman test is well known.

Its use in experimental physiology is reviewed by Kaplan (1962), who provides a wide variety of practical details relating to anaesthesia, care, surgical anatomy and procedures, standard values in blood, and use of the rabbit for the study of the various body systems. Since the thalidomide disasters the rabbit has been used very widely indeed in studies designed to detect a possible teratogenic effect or some other evidence of interference with the normal reproductive process, and this has expanded considerably its range of usage as a laboratory animal.

At the Huntingdon Research Centre, up to July 1964, well over 100 drugs had been screened in the rabbit, and the results of the first 67 of these were recorded by Worden and Harper (1964).

The ease with which the animals can be handled and fed is also a point in their favour, although the above-mentioned variations call for adjustment in housing, management and feeding.

There is available a considerable body of information on inheritance in the rabbit, in regard to which particular mention should be made of the work of Sawin and colleagues on inheritance of variations in reproductive phenomena and anatomical structure and of Nachtsheim on inherited abnormalities.

Much information is available also on diseases, both spontaneous and induced. Inherited variations, certain diseases, and many physiological processes are similar to those found in man and other domesticated animals.

A mimeographed bibliography *The Rabbit as used in Disease Research* compiled by Carlton C. Herman and published by the U.S. Department of the Interior is unfortunately out of print.

There appears to be little general agreement with regard to the most suitable types of rabbit for various purposes except that for venous injection and bleeding rabbits possessing large ears (lops, half-lops and large breeds) are often preferred.

Himalayan rabbits and New Zealand Whites are very suitable for most skin tests, for which unpigmented skins are almost essential. Table elsewhere in this chapter shows the characteristics of the common breeds available in Great Britain.

Few laboratories at the present time follow a policy of inbreeding, although the work of Lurie (1941) on the selective breeding of strains with enhanced or diminished susceptibility to tuberculosis, and other work of a similar nature, points to the value of an inbred strain for certain purposes.

Once the stock of rabbits in a laboratory has been established, the policy of closed rabbit-breeding should be followed in order to reduce the risk of introducing disease, whether or not inbreeding is used at the same time.

The Laboratory Animals Bureau, as it then was, of the Medical

Research Council conducted a survey during 1956, from which it was found that, of some 415 laboratories using animals, nearly half used the rabbit either alone or more usually as one among several species.

In that year approximately 38,000 rabbits were used, and from other evidence it would appear that this was a somewhat conservative estimate. Reference has already been made to the considerable use of the rabbit for teratogenic studies in Great Britain. In the 1956 census, the approximate allocation was as shown in Table elsewhere in this chapter.

Table 12.1: Utilization of Rabbits.

	Per cent
Diagnostic work	12
Research	53
Bioassay	18
Other purposes	19

Some 42 per cent of the rabbits were bred in laboratories where they were used. Some details of the physiology of the rabbit in health are given below in the section on diseases. Many normal values are recorded by Albritton (1952), Kaplan (1962) and Flynn (1965), to which reference should be made.

ACCOMMODATION

Accommodation of laboratory stock falls naturally into two parts: housing of breeding stock and housing for experimental procedures.

In both cases there are certain prime essentials. Each cage must be so constructed that it is comfortable for the occupant, is easily cleaned and disinfected, facilitates easy handling, experimental work and observation, excludes draughts, and can be kept dry.

Ventilation must be adequate, and for this reason it is preferable that cages should have open fronts. Outdoor housing is suitable for breeding-stock, but not for animals undergoing experiment or constant handling.

Some protection, such as a roof extending 3 or 4 ft (1 or 1½ m) in front of the cages, should be provided both to ensure that the rabbits themselves are protected and to afford good conditions for the attendant.

CAGES

At one time wood was used almost exclusively in the construction of cages, but for laboratory use metal cages are recommended except, possibly, in the case of outdoor breeding-hutches. An example of a typical wooden hutch is afforded by that designed by King Wilson (1935) which is a minimum of 2 ft deep and 2 ft 6 in long (61 and 76 cm); between two hutches is a hinged door that can be opened when a doe has a litter, thus affording double space.

For breeding, growing and resting stock, cages stacked in three tiers are suitable. Sufficient space should be allowed between rows of hutches for easy passage of a wheelbarrow or handcart. A minimum passage of 5 ft (1·5 m) is desirable.

All hutches, whether indoor or outdoor, must be raised a minimum of 9 in (23 cm) from the ground, and should not be stacked against walls. An air space of 91 to 12 in (25 to 30 cm) should be left between hutch back and wall.

In view of the range of breed sizes, hutches of many dimensions are used, but many are disgracefully small and prevent the occupants from taking reasonable exercise and from keeping away from the soiled corner in solid-floored cages.

Cramped quarters quickly lead to an increase in the incidence of disease and to poor growth. A suitable standard for the floor space of a breeding-hutch is 1 ft^2 per lb (0.23 m^2 per kg) of body-weight. The minimum size of any one hutch should be 2 ft (61 cm) square and 18 in (45 cnm) high.

Thus for the smaller breeds a floor space of 2 2·5 ft (61 × 76 cm) would be suitable, for medium-sized breeds 2 × 3 ft (61 × 91 cm), and for the larger breeds a minimum of 2 × 4 ft (61 × 122 cm). For growing stock and resting adults a slightly smaller area would suffice, but the larger area is always to be preferred, particularly as it is desirable that all hutches should be of standardized size and design in order to allow of easy management.

With regard to constructional details, the typical solid-floored hutch is made from a minimally 5-in (1½-cm) tongued and grooved boarding on a framework which should be of 2 2-in (5 × 5-cm) timber. A more satisfactory construction is the use of asbestos sheet over a framework of timber.

Metal hutches, which have many advantages, are always to be preferred indoors. They can be made with sheet metal (aluminium alloys are suitable) or woven or welded wire. The wire mesh for the walls and top can be of a mesh not exceeding 2 × 2 in (5 5 cm) the longer side being vertical.

The wire should be galvanized and of not less than 14 standard wire gauge. Perforated sheet, woven wire or welded wire panels of -8-in

(1½-cm) of slightly larger mash, made from a minimum of 12 s.w.g. has been found to be satisfactory for floors.

Woven wire is recommended in preference to perforated sheet. Individual wire pens mounted over trays on a trolley are excellent for indoor use. No bedding is required with self-cleaning mesh floors. For solid-floored hutches sawdust, hay or straw is satisfactory.

It is important that all bedding should be kept dry. Some wire cages are placed directly on a sawdust-filled tray and removed entirely from this when cleaning takes place.

Single infected rabbits can be kept in cages of the dustbin type described by Miles and Mitchell (1934). For the accommodation of rabbits for short periods a small wire cage measuring 18 a 18 in (45 × 45 cm) is useful.

Equipment

Removable nest-boxes are almost indispensable. For medium-sized breeds, boxes measuring 10 or 12 in (25 or 30 cm) high, the same in width, and 15 or 16 in (38 or 40 cm) long are suitable. Rather larger boxes are necessary for the biggest breeds, and smaller for the smaller breeds.

Nest-boxes should be totally enclosed except for the top half of one side. The lid and the remaining half of that side should be hinged in order to facilitate examination of the litters and cleaning. With the aid of insulated nest-boxes, breeding can be satisfactory at temperatures below zero. Hay overlying sawdust is the most suitable nesting material.

For hutches not equipped with automatic or semi-automatic feeding hoppers and watering devices, glazed earthenware bowls are suitable. Each feeding-bowl should have an inturned lip to prevent the animal from wasting food. Rabbits will drink readily from water bulbs which can be puchased or made cheaply.

Travelling-boxes for rabbits should be light but strong. Resin-bonded plywood is suitable. Measurements should be: height 12 to 14 in (30 to 35 cm); width: 8 to 12 in (20 to 30 cm); and length 14 to 18 in (35 to 45 cm); the smaller sizes in each case being for the smallest breeds and vice versa.

Ventilation should be through the lids and sides, with baffle boards to prevent any direct draught on the animal. The boxes should be waterproof.

Blount (1945) provides an illustrated account of the successful use of the Milanson Hygienic Hutch Unit which consists of a solid platform with a small hole through which the droppings and urine fall to a tray below. Apparently it is not difficult to train rabbits to use the unit by sprinkling the platform with a small quantity of soiled bedding.

Blount records that in some cases in which it was used, all the urine and about 80 per cent of the faecal pellets were collected in the trays. A modification of this unit can be easily manufactured from sheet metal.

A tray 12 in (30 cm) square or larger, with a depth of 1½ or 2 in (4 or 5 cm) is covered with a removable wire grid of ½-in (1½-cm) mesh. Most of the faeces are collected in this tray when placed in the appropriate corner of the hutch. It can then be emptied regularly.

BREEDING

The rabbit is particularly suitable for the study of problems concerned with reproductive physiology, and consequently there is an abundant literature on this species. In this chapter emphasis is placed on those aspects which have a direct bearing on breeding practices.

In the rabbit there is no oestrous cycle as in rodents, but some authors have suggested that a certain rhythm exists in sexual activity. Under favourable conditions, sexually mature does remain on heat for long periods during which Graafian follicles are continuously developing and regressing in such a way that approximately constant numbers are available for ovulation.

At certain times, for example during lactation or if nutrition is poor or when moulting, does may show no desire to mate, owing to a suppression of follicular activity.

How can oestrus be recognized? The vaginal-smear technique is not satisfactory. The appearance of the vulva offers a good indication, as it tends to enlarge and become purplish in colour during oestrus. However, it is not an infallible guide because some does will not mate at this time, whereas others mate even when the vulva is small and pale.

Owing to the apparent divergence between physiological and behavioural oestrus, it is worth while observing how the doe behaves when placed with an active male. If the doe refuses to accept the male it is advisable to try her with a second one.

It is particularly noteworthy how a doe's behaviour may vary with different males. Ovulation is generally non-spontaneous and normally occurs only after mating. However, under certain conditions conducive to intense sexual excitement, ovulation apparently occurs spontaneously, even in does occupying separate cages.

Variations exist in both the time of onset and duration of the ovulation process in individual does; the limits are from about ten to thirteen hours. The number of ovulations is related to bodysize. Some does (sometimes 25 per cent or more) fail to ovulate after mating, probably owing to a deficiency of luteinizing hormone (LH) in their pituitary gland.

Ovulationfailure is extremely uncommon after an intravenous injection of LH (25 Lu.), the use of which will consequently improve fertility. Induction of ovulation by sterile mating or LH injection leads to pseudopregnancy, which lasts approximately seventeen days.

During this period corpora lutea develop and the uterus and mammary glands show changes characteristic of early pregnancy. Nest-building activity is displayed at the end of pseudo pregnancy, when the corpora lutea regress.

Does may accept the male at any time during pregnancy or pseudo pregnancy, except for a brief period from 33 to 39 hours after the first mating, but ovulation does not follow when active corpora lutea are present.

However, ovulation can be induced at these times by LH treatment. In such cases fertilization may occur, provided the second ovulation takes place either within two or three days of the first, or towards the end of pseudopregnancy. Fertilization does not occur when active corpora lutea are present either because sperm-transport is interfered with, or because sperm-capacitation is inhibited in the progestational uterus.

Mating is best arranged in the male's quarters: if moved elsewhere bucks often show more interest in their new surroundings than in the doe. The characteristic sequence of mounting with exploratory pelvic thrusts culminating in copulation and orgasm takes only a few seconds.

Under good conditions 20 does can be mated in 30 minutes. Not infrequently does, although oestrous, may not permit mounting, especially if the male is young and inexperienced. In such cases, the practice of forced mating should be adopted.

The doe is restrained with one hand whilst the rear quarters and tail are raised with the other hand. When the buck mounts, the doe, if oestrous, will show lordosis. There are no hard-and-fast rules concerning the frequency of use of males, which show wide individual variations in response.

It has proved possible to ejaculate bucks as frequently as once a day for as long as ten months without impairing their libido, semenproduction, or fertility. In exhaustion tests, as many as 20 collections have been made over a 10-hour period. Over limited periods it may be permissible to exploit males for semen-production, but excessive use is not generally recommended.

A simple test for male fertility consists in obtaining a smear from the vulva immediately after copulation when semen is present. Microscopic examination should reveal the presence of actively motile spermatozoa, but this *per se* does not provide conclusive proof of fertility.

Ideally, a more critical analysis of male fertility should include a thorough examination of one or more semen samples collected with an artificial vagina, combined with a small series of test matings.

There are two important diseases, viz. syphilis or vent disease and snuffles, that can be transmitted at the time of mating. Care should be taken to avoid bringing infected animals of either sex into contact with healthy stock.

At copulation, upwards of 200 million sperm may enter the vagina, but probably no more than 1,000 reach the site of fertilization. Sperm movement in the female genital tract is a relatively slow process in the rabbit, requiring 3 to 5 hours before sufficient sperm enter the fallopian tubes to ensure maximal fertilization.

During their migration through the female tract, sperm undergo a process termed 'capacitation' which prepares them for penetration of the egg membranes. The nature of the capacitation process, which takes 6 to 10 hours to reach completion, is still obscure.

The fertilization process, which takes place in the fallopian tube at the junction of the ampulla and isthmus, begins shortly after ovulation. The first cleavage division, giving rise to the 2-cell stage, is observed 22 to 25 hours after mating and thereafter further divisions occur at frequent intervals, making the rabbit egg one of the fastest-developing in mammalia.

After 72 hours, when entry into the uterus takes place, blastocyst-formation is proceeding. Implantation begins late on the seventh day, by which time the blastocysts measure 5 mm in diameter.

The phenomenon of delayed implantation is unknown in the rabbit. Mating post partum, when oestrus is pronounced, initiates pregnancy, the outcome of which depends upon the number of young suckled. With small numbers, a pregnancy of normal duration may be superimposed upon the lactation.

However, if the number of sucklings exceeds a certain number (which varies with breed) pregnancy fails about the fifth day, apparently owing to regression of the corpora lutea.

Upwards of 25 per cent of eggs shed perish before term, the mortality being distributed in a characteristic manner—about 3 to 10 per cent before implantation and 20 per cent or more after implantation, with peak losses occurring from day 8 to day 17.

Pregnancy fails completely in about 5 per cent of does, which may, however, subsequently bear litters. Abnormalities of the genital tract such as might account for reproductive disturbances appear to be very rare.

However, in aged does a rather heavy incidence of uterine abnormalities may occur. So far no therapeutic measures are available for reducing prenatal mortality. Embryonic mortality may be induced experimentally by the injection of oestrogens, which terminates pregnancy, or surgically, in which case it is possible to eliminate selected embryos.

Using the technique of abdominal palpation, pregnancy can be diagnosed as early as 9 days after mating, when the uterine swellings measure only 12 min. As the swellings increase in size, reaching 20 mm at 13 days, pregnancy-diagnosis becomes easier.

At the Animal Research Station, Cambridge, does are palpated routinely on the tenth day after mating. In the few doubtful cases that arise, the examination is repeated some days later. With experience the same technique can be used to estimate the stage of pregnancy in does whose breeding history is unknown.

The technique is particularly useful in following the course of pregnancy in does treated experimentally with drugs, whose action may interfere with pregnancy.

Early pregnancy-diagnosis is important because it facilitates better management of the pregnant females and at the same time permits early remating of non-pregnant does. Test mating is not a reliable procedure for diagnosing pregnancy.

Usually gestation lasts 31 to 32 days, but there is a tendency for the smallest litters to be carried longer. In a group of 35 albino does, in which both the times of mating and parturition were known, the mean duration of gestation was 31·0 ± 0·8 days with a mean littersize of 6·3.

Parturition generally occurs during the early hours of the morning and for this reason it usually passes unobserved. Cross's (1958) observations on induced labour in full-term does showed that the whole process took only 7 to 20 minutes. In untreated does parturition is normally completed within 30 minutes.

Occasionally parturition is split, i.e. part of the litter is born several hours, even a day or more, before the remainder. Routine palpation of does postpartum, which is strongly recommended, will indicate any does having retained foetuses. Such does may be successfully treated with oxytocin* to induce labour.

This also applies to does which have not littered by the thirtyfourth day, when the chances of parturition occurring normally decrease. Young retained *in utero* beyond the thirty-fifth day die, and if they are not expelled the uterus becomes fouled with decaying tissue, which prevents further pregnancy.

Each foetus is expelled together with its placenta, which is almost always immediately eaten by the doe. The young, after being cleaned, move to the teats and suckle whilst the birth of their siblings proceeds. Curdled milk may be found in the stomachs of young removed from the nest shortly after birth.

Sometimes does, more especially primiparae, mutilate part or all of their litters. When cannibalism occurs on more than two occasions, offending females should be culled. Similarly does that do not prepare adequate nests, but scatter their newborn young about the cage, are hardly worth retaining as breeding-stock.

In 62 natural parturitions of primiparae, which produced 490 young, only 5 young were stillborn while 11 others died shortly after birth. In superovulated does the incidence of stillbirths is significantly greater, reaching 25 per cent or more.

Abortion is extremely uncommon; it occurs only after the twentieth day of pregnancy. If the entire litter dies before day 20 elimination takes place by resorption, which is very rapid.

Contrary to general opinion, sexing newborn young is not difficult, providing the technique is properly understood. It depends upon differences shown when slight pressure is applied to the external genitalia causing an eversion of the rudimentary penis or vulva.

In experienced hands the technique is very accurate and rapid. Before examining the litter it is advisable to remove the mother; she should only be returned after the young have been replaced in the nest.

The two sexes are produced with equal frequency. In 1,024 litters containing 9,854 young there were 4,928 males and 4,926 females, giving a sex ratio of 100·04. It must be emphasized that when only a few litters are considered the sex ratio may show wide variations. A claim to have controlled the sex ratio experimentally was made by Gordon (1957) but so far it remains unconfirmed.

Sexual development in the doe is influenced by several factors, including breed, nutrition and season of birth. There is a tendency for the smaller breeds to reach sexual maturity earlier than the larger breeds, and a high plane of nutrition also hastens the onset of sexual function. It is not possible to state precisely when does should be mated for the first time,

since so many factors are involved. Under good conditions most breeds will be ready from six to eight months. To delay mating too long is not good practice. One method that has proved its value consists in injecting virgin does with LH at five or six months of age to induce ovulation and pseudo pregnancy.

About the twentieth day after injection they are ready for mating. This technique offers two advantages; it permits synchronization of sexual function in groups of does, and at the same time the genital tract is stimulated without the extra load of pregnancy.

Breeding does should produce four litters annually. If the young are weaned at seven weeks of age, an interval of ten days will be available between weaning and mating. Little information is available on the culling of breeding does.

In large breeds a lifetime production of 80 to 100 young over three years constitutes a very satisfactory performance. The smaller breeds, in which litters are smaller, will produce fewer young. In older does, breeding troubles appear to be mainly associated with uterine abnormalities.

Superovulation

Rabbits are particularly suitable for the application of superovulation techniques involving treatment with anterior pituitary preparations, rich in follicle-stimulating hormone. Processed pregnant mares' serum has also been used successfully, but trials with the raw serum have been disappointing.

Anterior pituitary preparations, usually of horse or sheep origin, are injected subcutaneously in aqueous suspension either twice daily for three days, or once daily for five days. Twelve to twenty-four hours later an ovulating injection, 25 i.u. of LH, is given intravenously.

In adult rabbits mating alone is sufficient to induce ovulation of the extra follicles. The best response is observed in does aged 6 to 12 months, which may yield an average of more than 30 eggs each, with individuals giving up to 100 eggs. In aged females the response is diminished.

Superovulation can also be induced in immature females, in some strains as early as 12 weeks of age. Eggs obtained from superovulated females are quite viable, but when immature animals are used the proportion of abnormal eggs may be slightly increased.

Artificial Insemination and Egg Transfer

Artificial insemination is easily carried out and gives results at least as good as those obtained with natural mating, over which it has several advantages. Adams (1961) has described the methods used in semen collection, preparation of the does, and deposition of semen into the female genital tract. For a review of the literature concerning A.I..

Egg-transfer was first performed in 1890 by Heape. More recently much fundamental work, summarized by Austin (1961), has been done on various aspects of egg-transfer, particularly in the rabbit. The technique now constitutes an important research tool.

Selection

There are certain primary requirements for all breeding stock. Sawin and Curran (1952) have shown that the development of strains with good reproductive efficiency, which is essential to success, involves rigorous selection for six major factors: fertility, fecundity, milk-production, maternal behaviour, growth rate, and viability.

Although environmental factors affect these characteristics, these are basically inherited. Records of them should be incorporated in the recording system used in the animal house. The only characteristic which cannot be determined directly is milk-production.

An index of this characteristic can be obtained by using the weight of the litter at 21 days, or when the youngsters first leave the nest if they do so earlier. Youngsters, particularly of the larger types, leaving the nest before the age of about 21 days do so mainly through hunger, thus indicating poor milkyield in the doe.

Management

Does should not be permitted to produce more than four litters a year but, conversely, inadequate breeding is unsatisfactory, from both an economic and a breeding point of view.

The litter-size should be restricted to six or seven youngsters, but smaller litters are equally undesirable. Nest-boxes are most useful in that they allow easy examination of the litter, prevent chilling, and to some extent prevent too early emergence of youngsters from the nest.

Nestboxes should be placed in the hutches of pregnant does one week before parturition, and lined adequately with sawdust and hay or straw chopped into 6-in (15-cm) lengths. Spare supplies of fur from previous nest-linings may be added after sterilization to those nests which are poorly lined, the day prior to parturition.

Fostering of some youngsters from large litters to smaller litters can be accomplished up to three weeks after birth, although it is preferable to foster within a few hours of birth. Again, some litters may be amalgamated and the does without litters remated immediately after fostering.

Youngsters differing in age by as much as three days may be fostered, but it is preferable in this case to foster into her litter younger rabbits than those of the foster doe. No special precautions need be taken when youngsters are fostered on to a proven doe, other than giving her some pieces of food.

When a doe is unknown it is wise, when fostering, to remove her for half an hour or, if the youngsters are up to three weeks of age, for several hours

MANAGEMENT AND HANDLING

Weaning

The youngsters are usually weaned by the age of eight weeks. Slightly earlier weaning (minimum six weeks) may be practised but it is not desirable. At eight weeks the youngsters are best kept in pairs or colonies.

Separation of Sexes

All growing stock should be separated into sexed groups or pairs by three months of age. At this time some well-developed males will fight and may attempt copulation.

Marking

Permanent methods of marking include the use of rings, tattooing in the ear, and earnotching. The latter is the least satisfactory. Rings (which may be obtained in nine different sizes from the British Rabbit Council, Farnborough, Hants) are placed over the hock joints before the rabbits are eight to twelve weeks of age, according to breed.

Tattooing instruments may be obtained from surgical-instrument makers. The inside of the ear should first be cleaned with spirit, the tattoo marks made with a punch or needle, and ink rubbed in. Ear tags similar to those used for wing-marking chicks are sometimes satisfactory. Very young rabbits (from day-old) may be marked with an aqueous solution of acriflavine or gentian blue.

Handling

All handling should be firm but gentle. Provided that the animal feels secure it is less likely to struggle. The most satisfactory method of lifting is to place one hand under the body to take the weight, the animal being steadied with the other hand grasping the ears as close as possible to the head.

When rabbits have been subjected to injection in the ear or to bleeding, they can be steadied by grasping the loose skin over the shoulders instead of the ears. Young rabbits can be lifted by grasping firmly across the loins, the fingers on one side, the thumb on the other.

Care must be taken when using this method to avoid internal injury. An alternative method of lifting, particularly used for lop-eared rabbits, is to grasp them firmly around the abdomen with one hand on each side.

This method also must be used with care, particularly when handling pregnant does. Rabbits should always be placed on a non-slippery surface to prevent fright, which is easily caused by slipping. A sacking-covered table is very satisfactory.

Restraint During Bleeding

Smith (1931) illustrates a method of applying restraint during bleeding. Bleeding-boxes are not too satisfactory. Thorp (1944) described a satisfactory box for holding rabbits for minor operations and bleeding. For good restraint, a rabbit may be securely wrapped in a towel or sack.

Routine Inspections

Each animal should be inspected regularly and thoroughly, particularly just before mating. Over-long nails should be clipped to within 4 in (6·4 mm) of the quick. Rabbits shaking their heads and continually scratching at the ears should be immediately examined for ear mange. During routine examinations, an assistant to hold the rabbit while the genital organs are carefully examined makes thorough examination easy.

Feeding and Watering of Experimental Stock

Experimental stock should not be full-fed immediately before operation or bleeding. Any animal which has been bled should be allowed free access to water.

When large volumes of blood have been taken, intraperitoneal injection of saline in volume equal to that of the blood taken is good practice, especially in the case of rabbits which are unaccustomed to being provided with water. The addition of a little alcohol to their water may encourage rabbits to drink.

Young Stock

The most critical age for the young rabbit is at birth, but there is a second critical stage at from one to three months, unless the animal has been provided with an adequate allowance of concentrated food. On a poor diet it becomes highly susceptible to various disorders during this period of maximum growth.

It is therefore especially important to provide adequate nutrition (a diet low in fibre, high in proteins, and containing adequate minerals), space for exercise, dry hutches, and good hygiene.

Records and Recording

Suitable records are essential if successful breeding of stock is to be carried out. An excellent system for use in the rabbitry is that described by Sawin (1950).

Hand Rearing

Hills and Macdonald (1956) of the Animal House and Department of Physiology, Guy's Hospital Medical School, London, S.E. I., have reported upon the successful hand rearing of rabbits. Their account in *Nature* is repeated here almost verbatim with due acknowledgements.

Rabbits that were completely free of coccidial infection were desired for experimental purposes. It appeared to be impossible to obtain such rabbits by normal breeding methods, as infection, usually very mild, occurred during contact with the doe in the nest.

It was therefore decided to hand-rear rabbits which had been removed from the doe at birth. The expression of milk from the doe was found to be impracticable, so cow's milk was used.

The protein content of rabbit milk is approximately 10 per cent, whereas that of the cow is 4 per cent, and therefore, in order to increase the protein, calcium caseinate was added at the level of 15 g for 10½ (284 g) of cow's milk.

After several attempts it was found that this mixture could be used until the animals were seven days old, when it was necessary to increase the amount of calcium caseinate to 17 g and at fourteen days old to 20 g. This level was maintained until the young rabbits were able to eat fresh green food when about four weeks old, at which time they were weaned.

The weight curves for the hand-reared animals and their litter-mates still with the doe showed that at about eight days of age there was little difference between the two groups, that at eighteen days of age the naturallyreared animals were about 40 g heavier, and that at twenty-two days of age they were about 6 g heavier. The calcium caseinate did not dissolve freely in the cow's milk unless the whole was beaten with an egg whisk.

This mixture would keep for several days in a refrigerator. The milk mixture was warmed by placing it in the feeding bottles and immersing them in a cup containing warm water. As delivery of the milk was not satisfactory with a pipette, feeding bottles with rubber teats as used for dolls were obtained.

After a hole had been made in the end of the teat, the only difficulty encountered was the over-anxiousness of the young rabbit to obtain its milk, and a little practice was necessary in order to insert the teat satisfactorily between the tongue and upper jaw of the animal.

The rapidity with which the milk was consumed did not appear to affect the animal adversely. Sufficient was judged to have been given when it would take no more despite a rest of a few minutes.

The feeds were given approximately every three hours, the last feed being at midnight and the first at 5 to 6 a.m. Theanimalswere transported in a box lined with hay and cottonwool and care was taken to ensure that they were at all times kept warm.

One danger that was met was the accidental inhalation of some of the milk, which caused a fatal pneumonia. Another hazard was an enteritis which was successfully treated with intra-muscular injections of chloramphenicol.

NUTRITION

For a detailed review of the literature, reference should be made to a publication of the Commonwealth Bureau of Animal Nutrition.

Energy Requirements

Maintenance

The basal or standard energy requirements of rabbits of different weights have been measured. Requirements for 24 hours, in round figures. calculated from the equation of Lee (1939b), increase from 60 kcal for rabbits weighing 1 kg, by increments of 20 kcal per 0·5 kg, to 220 kcal for rabbits weighing 5 kg.

According to Brody (1935) for mammals in general the energy value of the total digestible nutrients for maintenance should be about twice the basal expenditure. On this basis the energy requirements of rabbits for maintenance range from 120 kcal at a body-weight of 1 kg, by increments of 40 kcal for each additional 0·5 kg body-weight, to 440 kcal at a bodyweight of 5 kg.

Values for energy of digested food (EDF), which is the heat of combustion of the food less the heat of combustion of the faeces, measured by calorimetry, are available for many individual feeds.

Values for other feeds may be calculated, by applying to the digested nutrients of the foods, appropriate factors given by Merrill and Watt (1955) for heats of combustion of protein, carbohydrate and fat.

It is therefore possible to estimate the energy of digested maintenance rations for comparison with theoretical requirements. This has been done for several recommended maintenance rations and most were found to provide somewhat more than the estimated requirement.

Reproduction

The energy-requirements of pregnant does and breeding bucks have not been established. Recommended increases are usually expressed as a percentage of maintenance allowances. Scandinavian authorities recommend increases of from 5 to 40 per cent for both does and bucks. The National Research Council, Committee on Animal Nutrition,

U.S.A. (1954), gives the same allowance for pregnant does as for maintenance of does and bucks of the same weight. It appears that the less generous the allowance for maintenance the greater is the increase recommended for reproduction.

Most of rabbits' foetal growth takes place in the last third of the period of gestation and birth-weight of the young is affected by the diet of the doe during the period of rapid foetal growth. Sandford (1957) recommends one and one-third times maintenance at the start of pregnancy, rising to twice maintenance at the end.

Lactation

Estimates have been made of the additional energy of digested food required for peak milk-production by does, based on available information on yield and energy value of milk and assuming that the doe converts food energy into milk energy as efficiently as a good cow.

They are in good agreement with recommended allowances for lactating does, viz. two or three times the maintenance allowance by the third or fourth week of lactation. Inclusion of the litter's needs of solid food in the recommended allowances for a doe in late lactation raises these to four or more times the maintenance allowance.

Growth

Axelsson (1949a) gave standards for energy requirements in terms of metabolizable energy for weight-gain of growing rabbits at different ages. Axelsson's standards, converted to energy of digested food, together

Table 12.2: Estimated energy requirements for growth of rabbits of medium-sized breeds.

Age (weeks)	8	10	12	14	16	18	20	22	24
Approx. weight (kg)	0·96	1·28	1·62	1·87	2·10	2·27	2·41	2·55	2·67
Weekly gain (g)	156	170	128	114	85	71	71	57	—
EDF (kcal per g gain)	4·2	5·1	5·9	6·8	7·6	8·5	9·4	10·2	11·1

with average weights and gains in weight of rabbits of medium-sized breeds as given by Sandford (1957), are shown in Table elsewhere in this chapter.

Protein Requirements

Detailed recommended allowances of protein for maintenance, reproduction, lactation and growth are available from Scandinavia and the United States.

The former are given as amounts of digestible protein per food unit or per 1,000 kcal of metabolizable energy, and the latter as amounts of digestible protein per head per day and per cent of air-dry ration. The recommendations are summarized in Table elsewhere in this chapter.

Table 12.3: Recommended allowances of digestible protein.

	g per 1,000 kcal of metabolizing energy	*Per cent of air-dry ration*
Maintenance	30 to 36	12
Pregnancy	40 to 44	14
Lactation	48 to 52	20
Growth	46 or 48 at 2 to 3 months decreasing to	
Growth and fattening	36 to 38 in 8th or 9th month	16

It has been calculated from the National Research Council's examples of rations supplying recommended allowances of protein that the N.R.C. allowances in terms of grams of protein per 1,000 kcal of energy of digested food are 46 for maintenance, 58 for growth, 57 for pregnancy, 63 for lactation and 64 for growth and fattening.

Even allowing for the fact that Axelsson's allowances are related to metabolizable energy the American allowances are clearly much higher than the Scandinavian. The little experimental evidence available suggests that, at least for maintenance, growth and fattening, the N.R.C. allowances are excessive.

The proportion of energy derived from protein, expressed as nutritive ratio, is 1:2·9 for the N.R.C. maintenance diet, but studies of weight-maintenance and nitrogen balance suggest that a nutritive ratio of 1:6 or 1:8 is adequate.

According to the N.R.C. report for normal growth and fattening of rabbits between 4 and 7 lb (1·8 and 3·2 kg) live-weight, the requirement of digestible protein is 16 per cent of the air-dry ration. The expected average daily gain is said to be 1·2 oz (34 g) per head. Such an allowance of protein would seen to be excessive according to results of feeding trials made at Cornell University.

New Zealand White rabbits, weighing about 3 lb (1·4 kg) at the beginning and about 6 lb (2·7 kg) at the end of the trials, were given pelleted diets containing 12 to 19 per cent of crude protein in the air-dry diet. As judged by weekly weight-gain and efficiency of food-utilization the requirement of crude protein was between 13 and 14 per cent of the air-dry ration.

A digestibility study showed that one of the diets which had 14 per cent of crude protein contained 9·4 per cent of digestible crude protein. With that amount of protein, average weekly gains equalled or exceeded the average expected gain postulated in the N.R.C. report.

Rabbits given a diet containing 12 per cent of crude protein ate less, gained less, and utilized food less efficiently than rabbits given diets with 13 per cent or more of protein. In Denmark good growth and fattening was reported in rabbits at a progeny-testing station where the two breeds principally represented were White Landrace and French Voedder, of average weight about 3 and 4 kg respectively at 22 weeks.

It has been calculated that the Danish fodder plan provided 39 g of digestible crude protein per 1,000 kcal of energy of digested food and a nutritive ratio of 1 :3·4 or 1 :3·5 compared with 64 g per 1,000 kcal and a nutritive ratio of 1 :1·8 by the N.R.C. diet for growth and fattening.

Fat Requirements

It may be seen from the table of composition of rabbit feeds given by Aitken and Wilson (1962) that the feeds commonly included in dry rations contain from about 1 to 8 per cent crude fat. The effects on growth of rabbits of diets containing different levels of fat have been studied only with purified diets.

Wooley and Mickelsen were unable to evaluate the effect of fat content of diet on growth. Thacker's results suggested that in purified diets 10 to 25 per cent of fat was better than 5 per cent. Rabbits given 10 per cent or more of fat ate more feed and made better live-weight gains than those given 5 per cent of fat.

Fibre Requirements

A review of the literature yields little evidence likely to be helpful in the choice of fibre content of diets for rabbits. The best guide would seem to be the fibre content of diets upon which rabbits thrive. It has been calculated that the fibre content of commonly recommended diets ranges from about 10 to 16 per cent of dry matter for does with litters and about 15 to 26 per cent for other rabbits. A pelleted feed which is said to be satisfactory for all classes of rabbits contains about 13 per cent.

Vitamin Requirements

Vitamin A

Rabbits need vitamin A for normal growth, development and reproduction. Under natural conditions they derive their vitamin A from green food in the form of the carotene precursors. Good hay and dried green food such as grass meal and alfalfa retain appreciable amounts of carotene and are the principal sources of the vitamin in commonly prescribed winter rations and complete pelleted feeds.

Diets made up of cereals such as white maize, oats, wheat, oil-seed meals, meat meal, tankage, beet pulp, and dried yeast contain no detectable amount of carotene. Studies of the effects of vitamin-A deprivation were reviewed by Aitken and Wilson (1962).

Young rabbits deprived of the vitamin show the usual signs of vitamin-A deficiency which include xerophthalmia, keratinization of mucous membranes, inco-ordination, retarded growth, and extensive degeneration of the nervous system.

Fully-grown does given diets low in vitamin A may show no sign of acute vitaminA deficiency, but concentrations of the vitamin in the plasma and liver fall, and reproduction is adversely affected. Fewer living foetuses and more resorption sites are found towards the end of gestation than in normal does, and young which survive may show xerophthalmia and hydrocephalus.

Xerophthalmia may appear also in the does after parturition. There is evidence that 1,000 to 3,000 μg of carotene daily prevents the appearance of signs of deficiency and as little as 50 Itg of carotene daily was found to be protective.

It has been calculated that a winter ration containing good-quality clover or mixed hay with some alfalfa meal or yellow maize in the concentrate mixture would provide for growing rabbits about 3,000 to 4,000 μg of carotene daily, and a ration with alfalfa hay and alfalfa meal probably provides more than twice that amount.

It is possible that intakes of carotene would be much lower with hay of poor quality, and it is perhaps wise to add a supplement of vitamin A to winter rations of doubtful potency. Cod-liver oil is sometimes given as a source of vitamin A. The need for caution in the use of cod-liver oil will be discussed below.

Vitamin D

Rickets has been produced experimentally in young rabbits but rarely occurs spontaneously. Requirements of vitamin D are not known.

The possible need for vitamin-D supplements to rabbit rations was studied by Jari. A diet of alfalfa meal, dried sugarbeet pulp and concentrates, with or without fish oil as an additional source of vitamin D, was given to does during pregnancy and lactation and to their young after weaning.

The additional vitamin D made no difference to growth of young. X-ray photographs of the tibia showed slower calcification in the young given no vitamin-D supplement, but at twelve weeks in one experiment and eight weeks in another there was no difference in calcification with or without the supplement.

In one experiment the Ca : P ratio was about 1·4 : 1, in the other it was 9·1 : 1. In this study substantial amounts of alfalfa meal were eaten daily. Different results might be obtained with diets in which the roughage was a poor source of vitamin D. To be on the safe side some breeders include a little cod-liver oil in the diet.

Vitamin E

Evidence that vitamin-E deficiency adversely affects reproduction in the rabbit is slender. The most obvious manifestation of such deficiency in the rabbit is muscular dystrophy in young animals, and estimates of the vitamin-E requirement are based on studies of the amount needed for prevention and cure of muscular dystrophy.

The condition is readily produced in the laboratory by giving diets containing excessive amounts of lard and cod-liver oil. The ill effects of such diets become more pronounced when the diet is allowed to stand before use. The deterioration can be prevented by the use of suitable stabilizers against oxidative rancidity.

It has been estimated that 0·2 to 0·4 mg of a-tocopherol per kg of body-weight is needed by rabbits getting a semi-purified diet containing 8 per cent of lard and 2 per cent of cod-liver oil. Less was needed when cod-liver oil was excluded. Choline deficiency increases the tocopherol requirement. Selenium has been found to have a beneficial effect on some forms of vitamin-E deficiency in other species, but not on muscular dystrophy of rabbits.

There have been reports of the occurrence of muscular dystrophy in rabbits reared on some pelleted feeds. Fats or fatty acids which are likely to become rancid should not be included in ready-made diets unless suitable stabilizers are included also. Cod-liver oil, if given, should be provided fresh daily, not mixed in the concentrate mixture for storage, and the amount should not exceed 1 per cent of the diet.

Vitamin K

Vitamin K does not seem to be required for normal growth of rabbits but is necessary for successful reproduction. Females deprived of it show no difference in blood prothrombin from those to which it is given, but they abort, and placentae from aborted foetuses show haemorrhage in the decidual plates. Vitamin K prevents the abortion.

Vitamin C

A review of the relevant literature shows no evidence of a need for this vitamin by the rabbit.

Vitamin B

It is now well established that the rabbit can synthesize vitamins of the B complex. The soft faeces of the rabbit are rich in vitamins B, and since those faeces are eaten by rabbits free to practise coprophagy the total turnover of those vitamins may be several times greater than the amount provided by the diet.

Indeed it would appear that the diet need not contain vitamin B_1, riboflavin, biotin, pantothenic acid, or vitamin B_{12}. With some diets there may be a need for nicotinic acid or its precursor tryptophan and there is some evidence of a need for vitamin B_5. The literature is reviewed by Aitken and Wilson (1962).

Choline Requirements

Experimental studies with purified diets have established the rabbit's need for choline. Deprivation of it causes failure of growth, liver-damage, and eventually muscular dystrophy. On the basis of growth-trials the requirement was estimated to be 0·13 per cent of a purified diet.

Mineral Requirements

Adverse effects of deprivation or low intake of some minerals have been demonstrated experimentally, and minimum requirements have been assessed from the response to increased intake.

The findings, although not necessarily applicable under different conditions (for instance with different basal diets and different forms of mineral supplement) give a rough yardstick with which to measure the adequacy of the mineral content of recommended diets. The average mineral content of some rations recommended by the N.R.C. was estimated by Aitken and Wilson (1962).

Calcium

The requirement of calcium is not known. Spontaneous vertebral fractures have been reported to occur in rabbits given a diet containing 0·2 per cent of calcium, and opacity of the eye lens, tetany, and a low blood-serum calcium have been observed in rabbits on a very low intake, 0·08 mg of calcium a day.

An intake of 30 to 60 Ing a day protected against lens opacity and tetany. Rations recommended by the N.R.C. provided an estimated daily intake of about 2 g daily (1 per cent of the ration) for normal growth and pregnancy and 2·5 g daily (0·5 per cent of the ration) for a clactating doe with a litter of seven. The amount of calcium excreted by a doe at the peak of lactation is about 1·3 g.

Phosphorus

Hay of low phosphorus-content from land deficient in this mineral has been used in ,studies of the phosphorus requirement for growth of rabbits. According to one such study, increasing the phosphorus content of the diet from 0·12 to 0·18 per cent by changing the quality of hay gave improved growth and calcification in New Zealand White rabbits.

In another study of New Zealand White rabbits graded supplements of dicalcium phosphate were given to increase the phosphorus content of the diet above 0·07 per cent; increasing it up to 0.17 percent gave significantly improved growth and calcification. There was a tendency for 0·22 per cent of phosphorus to give better results, but the difference was not significant.

A study of the smaller Dutch breed for which calcium phosphate was added to diets containing soya forage low in phosphorus, indicated that requirements for growth were met by 0·164 per cent of phosphorus in the dry matter of the diet. The estimated phosphorus-content of the N.R.C. ration for normal growth is 0·48 per cent.

Iron and Copper

Rabbit's milk, like cow's milk, contains little iron. The stores of iron in the liver of the young fall steeply from the very high levels at birth, and young rabbits at the end of the suckling period may be anaemic. Diets of cow's milk have been used to produce anaemia experimentally in rabbits.

The anaemia, which was hypochromic and microcytic, was not cured by iron alone or by copper alone, but in a few rabbits given both 2 mg of iron and 0·2 mg of copper daily there was a rapid regeneration of haemoglobin. Other signs of copper-deficiency in the rabbit are greying of black hair, loss of hair, dry scaly skin, and retarded growth.

The copper content of the liver falls during the suckling period, rises only slightly at about two months, and in the adult rabbit remains below that of newborn rabbits.

The optimum requirements of iron and copper have not been established, but it may be noted that the amounts which caused rapid haemoglobin degeneration in deprived rabbits and, in the case of copper, prevented changes in the hair and skin, were very small compared with the amounts provided by the N.R.C. diet for normal growth, viz. 36 mg of iron and 2·7 mg of copper daily.

The toxic effects of an excess of molybdenum in experimental diets, 0·1 per cent or more, may be counteracted by an increasing intake of copper, but we have seen no report of excess intake of molybdenum by rabbits on ordinary diets.

Manganese

When rabbits are weaned to a diet of cow's milk supplemented with iron and copper, gross bone deformity develops and growth is retarded unless a supplement of manganese is given. A daily supplement of 0·3 mg of manganese prevented obvious signs of deficiency, but storage of manganese in the liver, and growth, improved when the supplement was 1·0 mg.

A supplement of 4 mg gave better growth than did 1 mg in one experiment, but not in another, so the optimum intake was not clear. A study of the feeding of rabbits for three generations on a basal diet of hay and potatoes grown on soil deficient in manganese suggests that I rng of manganese is ample.

Over the three generations studied the mean daily intake of manganese of rabbits on the basal diet was 0·93 mg, comprising 0·5 mg for young rabbits and 1·1 mg for adults.

In growth, appearance of vitality, length or weight of bones, and manganese content of tissues there was no difference in any generation between groups given only the basal diet and groups given the basal diet supplemented with manganese.

The mean intake of manganese of the group given the supplement was 5.05 mg, comprising 3.8 mg for young and 9 mg for adult rabbits.

Roth calculated from the mean manganese content of the rabbit body that even from the diet low in manganese less than 1 per cent of the manganese was retained. The estimated amount of manganese in the N.R.C. diet for normal growth is 8 mg.

Cobalt

The average intake of cobalt from the N.R.C. ration for growing rabbits is calculated to be 23 μg. It has been shown, however, that growth and mortality in rabbits given a diet of milk and maize grown in a low-cobalt area were not significantly improved by the giving of a supplement of 4 μg of cobalt daily.

The mean cobalt intake from the diet was less than 0·1 μg and it was concluded that the rabbit requires less than this amount if it requires any.

Magnesium

Poor growth and hyperirritability leading to convulsions and death have been described in rabbits on a purified diet low in magnesium. From studies with graded supplements of magnesium it was concluded that the magnesium requirement of the rabbit is between 30 and 40 mg per 100 g of diet when the magnesium is given as sulphate. The average magnesium content of the N.R.C. diets is about 250 mg per 100 g.

Potassium

Muscular dystrophy has been produced in rabbits by restricting the potassium content of purified diets to 0·3 per cent or less. Trials with supplements of potassium show that maximum growth on purified diets containing 20 per cent of casein was obtained with between 0·6 and 0·9 per cent of potassium in the diet according to Hove and Herndon (1955) and with 0·8 per cent of potassium according to Wooley (19.54). The average potassium content of the N.R.C. diets is about 1·4 per cent.

Iodine

The iodine requirement is not known. Diets containing cod-liver oil or fishmeal should contain ample. We have seen no report of the occurrence of iodine deficiency in rabbits fed on all-vegetable diets without a supplement of iodine, so with the possible exception of rabbits reared on crops grown in an iodine-deficient area there would seem to be no need to supplement diets with this element.

Salt

Lack of an adequate supply of salt causes loss of water from the body, low rate of growth, and low milk-production, and the health and condition of animals is profoundly affected.

Utilization of food is impaired. Recorded salt-consumption of New Zealand White rabbits fed on concentrates and hay with a small allowance of green stuff and a salt lick was on average 0.1 g daily for bucks and 0·24 g daily for does with litters.

Rabbits given salty feed and offered unsalted water increase water consumption when salt intake is greater than about 0·3 g daily. In North America salt is commonly added at a level of from 0·25 to 0·5 per cent of the total ration or as 1·0 per cent of mash. Complete pellets made from plant foods usually include 0·5 or 1 per cent of salt.

In the absence of further information as to the salt-requirements of rabbits, it would probably be best to provide a salt brick, as is done for cattle and sheep.

Requirements of Water

More than two-thirds of the weight of the newborn rabbit, and about two-thirds of the adult, is water. Water is constantly lost from the body in the excreta and through the skin. A lactating doe may lose as much as 200 g daily in milk. Water lost must constantly be replaced.

Voluntary intake of drinkingwater by rabbits varies with the water-content of the food. Water intake per unit of body-weight varies with the environmental temperature and decreases with increasing age.

Records of the water intake of laboratory rabbits maintaining normal growth on a pelleted diet at an environmental temperature of 70°F (21°C) were kept by Cizek (1961). Mean values and standard deviations, based on 5,000 daily observations on each sex, were for males 235·2 ± 95·61 ml per day, or 104·1 ± 37·79 ml per kg per day, and for females 240·0 ± 103·82 ml per day, or 99·5 ± 38·57 ml per kg per day.

The rabbits were mainly of the Dutch breed. In the study made by Johnson *et al.* (1957) New Zealand White rabbits drank about 120 ml per kg at 70 days of age, the amount decreasing to about 64 ml at 340 days in an environmental temperature of 83°F (28°C). When the temperature was 48°F (9°C) corresponding figures were about 76 ml decreasing to 46 ml per kg.

Insufficient water intake retards growth and development. It is a wise precaution to keep laboratory rabbits plentifully supplied with drinking-water.

Feeding Practice

Almost any of the green foods, roots, hay, and concentrate feedingstuffs used for larger livestock may be given to rabbits. The digestibility of such feeds by rabbits has been studied by several workers and the results have been collected in tables in Swedish with English subtitles. A table of coefficients of digestibility of commonly-used rabbit foods is given by Aitken and Wilson (1962).

Fresh green foods and roots are digested most efficiently; in particular, the crude fibre is more efficiently digested than that of dry feeds such as hay, some cereals and oilseed meals.

Dry feeds, because of their much higher contents of crude nutrients, provide even when moderately well digested more total digestible nutrients per unit weight of food than do succulent foods, and are essential components of rations for breeding and growing rabbits when a high level of productivity is required.

The fodder plan used at the Danish rabbit-progeny testing station on Favrholm is shown in Table elsewhere in this chapter.

TABLE 12.4: DANISH FODDER PLAN.

Winter				*Summer*		
Body-weight (kg)	*Concen-trates (g)*	*Turnips (g)*	*Hay (g)*	*Concen-Crates (g)*	*Green feed (g)*	*Hay (g)*
1·0	30	100	50	30	125	30
1·5	40	150	50	35	150	30
2·0	56	200	50	49	200	30
2·5	72	250	60	63	250	30
3·0	80	300	60	70	300	30
3·5	88	350	60	77	350	30
4·0	96	400	70	84	400	30

In this plan the hay is mainly clover and mixed grass, and the green food mostly lucerne except in early summer when it is clover and mixed grass. The concentrates in winter consist of 75 per cent of pelleted feed, 22 per cent of equal parts of barley and oats, and 3 per cent of monosodium phosphate.

In summer the mixture consists of 57 per cent of pellets, 40 per cent of grain and 3 per cent of monosodium phosphate. The percentage composition of the pelleted mixture is ground barley 22, ground oats 12, ground wheat 10, wheat bran 10, wheat feed 10, soya-bean meal 10, yellow maize meal 8, lucerne meal 5, meat-and-bone meal 4, fishmeal 3, skimmed-milk powder 2, vitamins 2, minerals 1·5 and charcoal 0·5.

In large-scale rabbit-rearing in the United States succulent foods such as fresh greens and roots are left out altogether, and a dry diet of hay and concentrates is given for maintenance and production. Such diets involve the minimum of labour, especially if a suitable mixture of concentrates is obtainable in pelleted form, and are very successful if given with ample drinking-water. In the N.R.C. report on requirements of rabbits a pelleted diet used by Kellogg *et* al. (1949), with alfalfa hay, is given as an example of a diet suitable for rabbits. The ingredients are listed in Table elsewhere in this chapter.

Table 12.5:

American pelleted diet		***British Pelleted Diet SGI***	
Ingredients	***Per cent***	***Ingredients***	***Per cent***
Soya bean oil meal	18	Best white fishmeal (67) per cent protein	10
Linseed oil meal	4	Grassmeal (18 per cent protein)	20
Alfalfa (lucerne) meal	40	Bran	40
Wheat bran	15	Sussex ground oats	12
Barley grain, ground	18	Middlings	18
Oats grain, ground	4		
Salt	1		

A pelleted diet devised by Bruce and Parkes (1946) and known as 'Diet 18' was used for some time in the feeding of laboratory rabbits in Britain, but recently evidence has accumulated to the effect that its use may be attended by certain troubles, viz. loss of condition and reproductive failure and in some instances muscular dystrophy.

Short and Gammage devised a new pelleted diet, Diet SGI, to take the place of Diet 18 and found that condition and reproductive performance of rabbits on this diet, supplemented with hay, were better than with Diet 18.

This has been the experience also at the Rowett Research Institute, where Diet SGI supplemented with vitamins, and for pregnant and lactating does with vitamins and hay, has been much more satisfactory as stock diet for a colony of Chinchilla Giganta than was Diet 18. The vitamins, in a premix which is added to the diet at the rate of -1 cwt (25·4 kg) to 2 tons (2,032 kg) of Diet SGI. One cwt (50·8 kg) of the premix contains 16,000,000 I.U. of vitamin A, 4,000,000 LU. vitamin D_3, both stabilized, 32,000 mg of riboflavin, and 100,000 mg of vitamin E. The ingredients of the new pelleted diet are listed in Table elsewhere in this chapter.

The amounts of water, crude fibre, and digestible crude protein, and the energy of digested food of the two diets have been calculated by Aitken and Wilson (1962). Both diets contain about 10 per cent of water and about 12 per cent of crude fibre.

The American diet has the higher content of digestible crude protein, about 19 per cent compared with about 16 per cent in Diet SG I, and has the higher energy of digested food, about 300 kcal per 100 g as compared with about 260 kcal. Per kilocalorie of digested energy of food the amount of digestible crude protein is about the same in the two diets.

Recommendations as to the quantities of Diet SGI and the amounts of *hay,* if any, to be given to rabbits with it for maintenance, reproduction and growth have not yet been made. The American pelleted diet is recommended at a level of 0·25 lb (113 g) daily with 0·07 lb (32 g) of alfalfa hay at a live-weight of 4 lb (1·8 kg), increasing to 0·39 lb (180 g) of pellets with 0·08 lb (36 g) of alfalfa hay at a live-weight of 7 lb (3·2 kg). Those amounts are for New Zealand White rabbits gaining on average 1·2 oz (34 g) a day.

An example of a ration meeting the N.R.C. recommendations for normal growth of a rabbit of average weight 6·5 lb (3 kg), is alfalfa. hay 0·28 lb (126 g) with 0·16 lb (73 g) of a mixture of 4 parts of oats, 4 parts of barley and 2 parts of soya-bean-oil-meal pellets. For maintenance of a doe or buck of average weight 10 lb (4·5 kg), the ration is clover hay 022 lb (100 g) with 0·16 lb (73 g) of a mixture of equal parts of oats, wheat and linseed-oil meal.

A common method of supplying dry rations is to leave the hay in racks for the rabbits to eat to appetite and to supply the concentrate mixture in two or three feeds a day. Short (1960) has used with success an automatic rabbit battery with cages around which runs a rail carrying containers of food and water. The rabbits are on wire mesh, and excreta falling through to a sealed floor are swept away by devices attached to continuously-moving carriages.

Antibiotics are added to the food of most species of livestock to stimulate growth, but a review of the available evidence (Aitken and Wilson, 1962) suggests that antibiotics have no significant effect on weight-gain in the rabbit and are not associated with improved foodconversion except in experiments in which high mortality is reduced by an antibiotic.

Artificial Feeding of Baby Rabbits

See under 'Hand-rearing' above, and Hills and Macdonald (1956). However, Bernard (1962) has not found this diet and method of feeding satisfactory. Experiments now in progress suggest that the addition of vitamins to the diet, and restriction of the number of daily feeds to two, may be beneficial.

DISEASES

The whole subject is reviewed in detail, in its practical aspect, by Flynn (1965), to whose book reference should be made. As he points out in it, the healthy, well-nourished rabbit will not present any bony prominences over the shoulders, back or hind quarters, and the muscles over these areas should feet firm to the touch.

The forelegs should be straight and parallel to one another, and the hind-legs should be positioned well under the body. The portion of the hind-leg that rests on the cage floor is, of course, responsible for carrying much of the weight of the animal, and should be free from lesions. Fracture of a hind limb is common.

The external ears are mobile and can be moved independently of one another. The slightest unusual sound will cause the rabbit immediately to raise its head and erect its ears. When alarmed, animals of both sexes stamp the floor loudly with one or both hind feet.

Flynn (1965) notes that the pregnant doe appears to be less excitable than either younger animals or adult bucks, and appears to be much less bothered by outside noises and is altogether a more docile animal. The doe with newborn young may, however, be aggressive and strongly protective towards her offspring.

This protective attitude decreases as the young become more independent. The most obvious respiratory movements in the normal, quiet rabbit are to be seen on the sides of the abdomen, those of the chest wall being much less conspicuous.

The respiration rate in the normal adult animal is within a range of 38 to 65 per minute with an average of about 50. In the young rabbit, however, the rate is much higher, and in the infant it may exceed 100 respirations per minute. The volume of air taken in per breath by the adult rabbit is of the order of 20 ml.

During respiration the central flap between the nostrils lifts to show the obliquely elongated openings of the external nares. In the normal rabbit, the nares are dry and the fur adjacent to them is clean.

The skin of the normal rabbit is smooth, and the superficial lymph glands are inconspicuous. The fur is made up, in most breeds, of long guard hairs and an undercoat, and there are modified hairs on the under surface of the feet, and others forming the eyelashes and whiskers.

In Rex rabbits the long hairs are absent and the whiskers are poorly developed while the Angora has very long hairs. The moulting process, that usually occurs annually, tends to start on the fore parts of the shoulder and to proceed downwards across the flanks, the ventral parts of the body being the last to attain a new pelage.

Hair-chewing may occur when diets are low in roughage. The rectal temperature in a healthy animal is usually within the range of 101·5° to 104·2°F (38-5 to 40°C), the average being 103·1°F (39·5°C). As is well known in the laboratories where pyrogen assays are made, the rectal temperature may be modified by excitement, handling, or other environmental noises.

In the B.P. test for pyrogens temperatures are recorded on animals in a quiet room, which must have a temperature within 3° of the rabbits' living quarters, or in which they have been kept for at least eighteen hours before the test.

In practice in many laboratories, as at the Huntingdon Research Centre, the temperatures of the living and test quarters are identical, and the animals are maintained in this environment for up to two weeks before the test is made.

The process of refection has already been referred to and, as Flynn points out, the amount of faecal material passed during the day may vary considerably in the normal animal, and may even be twice as much on one day as on the next, even on a standard diet. With hay or other forms of roughage the quantity is increased.

The amount of urine passed per day is also somewhat variable and depends on the availability of water and green food. In very young animals it is colourless and does not contain any precipitates, but as soon as the rabbits consume solid or green food the urine changes in colour and a precipitate develops.

The colour may be lemon, straw, amber or red-brown, and the reaction is usually alkaline. The precipitate gives a positive reaction for albumen, which must therefore be accepted as a normal finding in the adult rabbit.

Coccidiosis

Perard (1925), Chapman (1929), Smetana (1933), Becker (1934) and Peck (1934) give useful information about this disease, which occurs in two forms, hepatic and intestinal.

Hepatic Type

Cause

Eimeria stiedae, a species of coccidium peculiar to the rabbit. It enters the bloodstream from the small intestine and attacks the columnar cells of the bile ducts of the liver, in which it completes its asexual cycle.

The sexual cycle involves the formation of the oocyst from the zygote and the accumulation of these bodies in the bile ducts produces the marked dilatation and thickening of the ducts which is characteristic of hepatic coccidiosis.

The oocysts are excreted into the gall bladder, thence reach the gut, and are excreted in the faeces. After an incubation period which varies with the conditions of temperature and humidity, the oocysts become infective and, upon ingestion by the same or another rabbit, the life-cycle is recommenced in the biliary epithelium.

Clinical symptoms

The severe form of the disease causes extensive enlargement of the liver by pressure and distortion of the bile ducts, and symptoms of unthriftiness and abdominal distension become apparent in infected rabbits two or three months after weaning. before which time this type of coccidiosis is rarely fatal.

Diagnosis

The disease is easily detected by microscopic examination of wet smears of the faeces for oocysts but unfortunately the oocyst count is not a reliable index of the severity of infestation, which can only be estimated by a post-mortem examination of the bile ducts.

Control measures

Therapeutic treatment of this type of coccidiosis will be described together with that of the intestinal type. Prophylactic measures aim at the elimination of the infective stage of the protozoan and include:

(1) the provision of sufficient floor-space per rabbit;

(2) timely weaning and separation of the litter from the parent doe;

(3) regular and frequent cleaning of the floor and sides of the hutch followed by disinfection with 10-per-cent commercial or household ammonia;

(4) adequate ventilation and the use of feeding-troughs and water-dishes of the unspillable type to avoid the conditions of temperature and humidity favourable for germination of oocysts (a fine-mesh, wire-netting floor is ideal for preventing the ingestion of droppings); and

(5) avoidance of feedingstuff (hay or greenstuffs) contaminated with the droppings of wild rabbits.

Intestinal Type

Cause

Eimeria perforans, another coccidial species peculiar to the rabbit, parasitizes the intestinal epithelium, causing rupture of the cells with haemorrhage in severe infestations. The life-cycle, apart from invasion of the blood-stream, is similar to that of the hepatic species. *E. magna is* another pathogenic species causing intestinal coccidiosis in rabbits.

Clinical symptoms

The severe and often rapidly fatal form of the disease affects suckling rabbits housed under conditions which are favourable to massive infestation. Such rabbits exhibit a ravenous appetite, become progressively emaciated, and may scour badly.

Diagnosis

The disease is detected in a similar manner to hepatic coccidiosis and while an oocyst count of the faeces is again an unreliable index of the severity of infection, the excretion of large numbers of schizonts is a more valuable indication. Post-mortem examination of the intestinal wall for the extent of the lesions is the best method of diagnosis.

Control measures

Those described for the hepatic type are equally applicable to intestinal coccidiosis.

Chemotherapy

The chemotherapeutic control of intestinal coccidiosis in rabbits now appears to be feasible. The administration of sulphadimidine (sulphadimethylpyrimidine or 'sulphamezathine') in the drinking-water has led to a striking diminution in the number of deaths from coccidiosis in a rabbitry containing 30 breeding does.

The death-rate from coccidiosis remained very high in untreated litters. The drug was given as a saturated watery solution. This was substituted for the drinking-water as soon as the doe had littered down. The treatment was given throughout the suckling period and was continued for the young stock after weaning until they reached the age of four to live months.

Treatment of the hepatic type with sulphadimidine is very effective provided it is started before the 10th day after infestation. Experimental results showed that a mash containing 1 per cent of the drug was efficacious in preventing deaths from hepatic coccidiosis when fed to rabbits which received numbers of oocysts sufficient to kill most untreated rabbits.

Treatment was begun 24 hours after the infective dose. At post-mortem examination the liver-weights of the treated rabbits were found to average one-twentieth of the body-weight as compared with one-ninth in the untreated ones which had died of the experimental disease.

Differential diagnosis

The oocyst of *E. stiedae* is distinguished from that of *E. petforans* by the greenish colour of its contents, and that of *E. magna* is larger than either, is ovoidal, and possesses a differential micropyle feature. Rutherford (1943) describes the morphology of four intestinal species.

Mucoid Enteritis

Within recent years this disease has ranked next in order of importance to coccidiosis and is probably more frequently fatal. It apparently affects rabbits of all ages but the deathrate varies from 100 per cent among suckling rabbits to about 33 per cent among weaned rabbits.

The incidence of mortality among adult rabbits is much less, and death generally occurs in affected does just before or after littering. For records of the occurrence of the disease in Great Britain see Muir (1943), Blount (1945). See also McKenney and Shillinger (1945).

Cause

The origin of this disease is unknown. Apart from the fact that it does not appear to be caused by bacteria, and is very probably not a virus disease, its aetiology remains speculative.

Clinical symptoms

The condition is often mistaken for coccidiosis, since the outstanding feature is generally marked emaciation, involving the musculature of the back and hind-quarters. There is no febrile reaction and affected rabbits may show a normal appetite up to the terminal stage of dullness, inappetence and general weakness. Post-mortem examination shows varying evidence of digestive disturbance.

Diagnosis

The disease is easily differentiated at autopsy from either of the types of coccidiosis by the absence of significant hepatic or intestinal lesions. Suckling rabbits, apart from signs of retarded growth-rate, usually show no macroscopic lesion of any kind but renal erosions can be detected in the majority of rabbits dying after weaning-age. An occasional rabbit surviving from an affected litter of which most died at suckling or weaning-age has died showing urinary incontinence and weakness of the hind-quarters after losing condition gradually for several months.

Control measures

The usual precautions of disinfection and isolation are unavailing. Experimental study of the disease has shown that affected does are liable to have a continued mortality among successive litters and that survivors from such litters have a similar breeding history.

From the breeding aspect it is best to ensure that breeding-stock is derived from rabbitries which are free from the disease.

Nutritional Muscular Dystrophy

Clinical symptoms of muscular dystrophy have been described in experimental rabbits fed on diets which were deficient in vitamin E, and have been ascribed to a loss of muscle creatine due to the excretion of abnormally large amounts in the urine.

Comparison of these symptoms with those of emaciation of the muscles of the back and hind quarters, observed in the condition described in the preceding five paragraphs, together with the abnormally high creatine/creatinine ratio detected in the urine of rabbits from a proportion of these outbreaks suggest that where no symptoms of digestive disturbance were seen the rabbits might well have been suffering from nutritional muscular dystrophy.

Cause

Insufficient vitamin E in the food or partial destruction of the vitamin-E content by cod-liver oil fed as a supplement.

Clinical Symptoms

Marked signs of muscular dystrophy in adult does soon after parturition and in their litters. In severe cases, the suckling rabbits are born with extreme dystrophic muscular lesions.

Diagnosis

Impaired weight-gain, or rapid loss of weight, with emaciation of the musculature of the back and hind quarters are characteristic of the disease, and stained sections of the affected muscles show necrotic changes in the fibres. An abnormally high ratio of creatine to creatinine in a 24-hour urine sample is diagnostic also.

Control measures

The disease may be prevented or cured by the administration by mouth of alphatocopherol in a daily dose of between 0·6 to 1·0 mg per kg body-weight. The vitamin preparation must be fed before other food (preferably on an empty stomach), as the antidystrophic activity of wheat-germ oil has been shown to be destroyed when the oil is incorporated in the basal diet.

Natural alpha-tocopherol succinate has also been recommended for vitamin-E therapy in the smaller furred animals in a daily dose of 100 i.u. for four or five days, followed by a maintenance dose of 25 i.u. per day.

Snuffles

The nasal discharge characteristic of this condition has given rise to its name. It is highly contagious, owing probably to dissemination of this discharge. A valuable account of it is given by Webster (1924).

Cause

Two bacterial species, *Pasteurella lepiseptica* and *Brucella bronchiseptica,* both members of the normal respiratory flora of the rabbit and occurring separately or together, are believed to be the cause of this disease.

Clinical symptoms

In the mild form the only constant symptoms are repeated bouts of sneezing, with or without a nasal discharge, the rabbit being otherwise normal and in good condition. The mortality associated with this type of the disease, generally ascribed to infection with *Brucella bronchiseptica,* is negligible, although the condition usually spreads to the entire rabbitry.

In the chronic type of snuffles the nasal discharge becomes very marked and gives rise to a characteristic snuffling noise. Affected rabbits progressively lose condition and die of a terminal pneumonia. The ultimate incidence of mortality is high.

The acute form of the disease, which is rapidly fatal, is characterized by a primary nasal discharge, the appearance of which is quickly followed by febrile symptoms. Death ensues so rapidly that the affected rabbit may appear in relatively good condition at postmortem examination.

Diagnosis

The mild form is recognized by symptoms of recurrentsn eezing with maintenance of good bodily condition and absence of deaths. The obvious nasal discharge with gradual loss of condition, resulting in a heavy death-rate, is indicative of the chronic type. The postmortem lesion of fibrinous pleuropneumonia is characteristic, and the causative bacterium is generally isolated in pure culture. The febrile symptoms of the acute form with accompanying nasal discharge are unmistakable. At autopsy, acute pleuropneumonia and/or peritonitis are the outstanding findings often with a supervening septicaemia.

Control measures

The strict eradication of infected and contact rabbits is the most reliable method of dealing with this disease, in view of its very contagious properties. Intravenous injection of a soluble sulphapyridine preparation usually cures the acute type, but a chronic infection persists. In the same way, intranasal insufflation with sulphapyridine powder seems merely to prolong the course of the chronic type of infection.

Pseudotuberculosis

This disease is quite commonly encountered in the domesticated rabbit and is essentially chronic. It is described by Will (1931) and by Seifried .

Cause

Pasteurella pseudotuberculosis rodentium, a bacterium commonly involved in diseases of wild rodents. The path of infection is by ingestion of material soiled by the excreta of infected mice or rats or of other infected rabbits.

Clinical symptoms

The main feature of the condition is a slow progressive loss of flesh with an ultimate resultant state of emaciation.

Diagnosis

The post-mortem picture consists of primary Gaseous nodules in the intestine, generally most marked in the caecum, with secondary irregular caseous lesions in the mesenteric lymphatic glands, the spleen and the liver. Joint lesions have been described in very advanced cases. The causative *Pasteurella* is easily differentiated from the snuffles *Pasleurella* by its motility, resistance to bile salts, and biochemical reactions.

Control measures

The incidence of this disease is reduced by prevention of contamination of foodstuffs and general measures of disinfection applied particularly to feeding and cleaning utensils. Clinically recognizable cases should be immediately killed.

Tuberculosis

This is a much rarer condition in the domesticated rabbit than is pseudotuberculosis, and usually only single isolated cases are met with.

Cause

Bovine and avian types of *Mycobacterium tuberculosis,* much the more frequently the former.

Clinical symptoms

Similar to those of pseudotuberculosis, with a more rapid course of infection. Lung lesions generally give rise to an increased respiratory rate in the later stages of the disease.

Diagnosis

The post-mortem features of the infection, whether of the bovine or of the avian type, are those of generalized tuberculosis, viz. multiple irregular necrotic areas in the lungs and liver, numerous spherical nodules of varying size in the spleen and kidneys, and caseous enlargement of the associated lymphatic glands.

Ziehl-neelsen's staining method demonstrates an abundance of acid-fast bacilli in films of the affected tissues. The intradermal tuberculin test is not so reliable in the rabbit as in the guinea-pig.

Control measures

Immediate slaughter of infected rabbits and incineration of the carcass is essential. The incidence of the disease, low as it is, might be reduced by avoiding milk which may contain tubercle bacilli and food contaminated by the droppings of poultry or other birds.

Necrobacillosis

This is a chronic disease which occurs sporadically amongst domesticated rabbits and is usually characterized by subcutaneous swellings irregularly distributed over the head and body.

Cause

Fusiformis necrophorus, a filamentous bacillus, which is a member of the skin flora of the healthy rabbit and enters the blood-stream by way of infected skin wounds or via the alimentary tract.

Clinical symptoms

The affected rabbit may appear in good condition in the early stages of the disease before the infection has progressed beyond the primary subcutaneous or internal necrotic lesion.

Thereafter, as the region of necrosis extends beneath the skin, with accumulation of greenish caseous material at the site of a skin wound very commonly sustained during a fight with another rabbit, the animal gradually loses flesh and its fur becomes ragged and dull.

The insidious spread of the infection by the subcutaneous route to other parts of the body and by the circulation to internal vital organs such as the liver, kidneys and lungs, where extensive necrotic areas are produced, results in ultimate emaciation and death.

Infection may spread to contact rabbits by the dissemination of the contents of a ruptured abscess with resultant contamination of skin wounds.

Diagnosis

The causal bacillus is readily demonstrated by microscopic and cultural methods in the caseous lesions under the skin and in the internal organs.

Control measures

All skin wounds should be cleansed and dressed with a reliable non-irritant antiseptic immediately upon detection. Any rabbit showing skin abscesses should be disposed of, as experience has proved that it is practically impossible to prevent, by surgical and antiseptic treatment, the subcutaneous spread of infection from an established primary skin lesion.

Rabbit Syphilis

This is a relatively common venereal disease of the domesticated rabbit. (McKenney and Shillinger, 1934; Seifried, 1937, pp. 81-96.)

Cause

Treponema cuniculi, a spirochaete resembling that which causes syphilis in man, but harmless to man and other domestic animals.

Clinical symptoms

The primary lesions appear in the form of papules up to the size of a pea on denuded areas of the skin of the external genitalia. Weeping ulcers up to a halfpenny in size follow the rupture of these vesicles and may spread to the hocks.

Secondary lesions often develop on the lips, nasal mucous membrane, and eyelids, involving the conjunctiva and cornea. Occasionally death follows an extension of ulceration, with generalization of infection, causing lesions in vital organs.

Generally, however, affected adult rabbits remain in moderately good condition, although young rabbits may show unthriftiness. Severe ulceration of the skin around the anus and genital orifice may cause diffrculty in passing faeces and urine.

Diagnosis

The causal micro-organism can be demonstrated microscopically in wet preparations of the exudate from the lesions by dark-ground methods and in dried films by Fontana's impregnation method.

Control Meaures

Incoming rabbits of both sexes should be carefully examined for evidence of the disease, especially when intended for breeding purposes. Cages and hutches should be disinfected with a blowlamp after housing affected rabbits, even though the latter may have recovered from the disease.

A single intramuscular injection of 025 to 035 g of neosalvarsan in 1·5 ml of distilled water constitutes an efficient cure for the disease. Simultaneous treatment of the ulcers with 25-per-cent calomel-lanolin ointment after cleansing with boric acid accelerates recovery. The disease also responds to treatment with an antibiotic such as penicillin.

Pyogenic Infections

These are of sporadic occurrence in domesticated rabbits, usually involving the salivary or mammary glands or the uterus.

Cause

Staphylococci and other pus-forming bacteria. *Fusiformis necrophorus* commonly occurs as a secondary invader.

Clinical Symptoms

A large, firm swelling, varying in size up to that of the fist, is observed between the lower jaws, or at the side if a salivary gland is involved. In the case of mastitis the teats and udder are inflamed, the mammary excretion is purulent, and the doe refuses to suckle her young. Pyometra is recognized by a chronic, purulent uterine discharge and a history of unsuccessful mating.

Diagnosis

All three conditions are characterized by evidence of a chronic local infection with suppuration.

Treatment

The abscess of the jaw, which usually shows little tendency to burst, should be incised and evacuated, and the cavity should be kept irrigated daily with a reliable antiseptic solution until healing by healthy granulation tissue occurs.

In mastitis, the pus should be removed daily from the udder by gentle massage, and the overlying skin should be treated with a mild counter-irritant ointment. For pyometra, frequent irrigation of the uterus and vagina with 5-per-cent hydrogen peroxide should be carried out.

Septicaemic Conditions

These generally occur during an epidemic outbreak leading up to acute infections, as soon as the causal bacterium has acquired increased virulence.

Cause

Pasteurella lepiseptica and *Salmonella typhi-murium* in young rabbits and *Pasteurella lepiseptica, Pasteurella pseudo tuberculosis rodeotium* and *non-haemolytic streptococci* in adult rabbits, particularly in does after parturition.

Clinical symptoms

Elevated temperature, accelerated respiration and loss of appetite.

Diagnosis

The causal bacterium can be isolated by cultural methods from the blood and internal organs.

Treatment

A course of intravenous injections of sulphapyridine solution in a dose estimated according to body-weight has had a curative effect in a few outbreaks of septicaemic infection.

The usual methods of isolation of affected rabbits, disinfection of contaminated hutches, feeding, and cleaning utensils and careful disposal of carcasses should be practised during and after an outbreak.

Pyogenic and septicaemic conditions have been successfully treated with antibiotics such as penicillin and terramycin by the injection of the recommended dose.

Metabolic Disorders

Pregnancy toxaemia

This is a fairly common disease causing mortality amongst does in the terminal stages of pregnancy; for a description, see Greene (1937-1938).

Cause

It is associated with endocrine imbalance arising from a disturbance of functions concerned in reproductive processes.

Clinical symptoms

A few days before parturition is due the doe shows a sudden onset of dullness and inappetance. Death follows within two or three days of the appearance of symptoms.

Diagnosis

The condition is recognized by the history and by the severe fatty changes observed in the livery at autopsy.

Treatment and control measures

These must remain undetermined until the aetiological factors have been ascertained. Prophylactic administration of glucose in the food has been recommended to prevent acetonaemia.

External Parasitic Conditions

Ringworm

Both types of this malady are of very rare occurrence in the laboratory rabbit, mice or rats being the usual source of the infection.

Cause

Various *Microsporon* species or *Trichophyton tonsurans*.

Clinical symptoms

One type is characterized by circular, raised greyish or yellowish crusts on the nose, face or ears and uncommonly on the feet and body, resulting in destruction of the hair-roots and follicles. The main feature of the other type is the appearance of circular, hairless spots on the fur with a covering of fine, pearl-grey, shiny scales.

Diagnosis

The branching filaments and characteristic spores of both genera of fungi can be detected microscopically in wet preparations of material from the lesions.

Control measures

Prevention of access of rodents to the rabbit-cages or hutches. An infected rabbit should be strictly isolated if treatment is undertaken; this consists of removing the crusts with a cleansing solution and dressing with a suitable fungicide such as diluted tincture of iodine, 10-per-cent salicylic acid, a 2-per-cent formalin salve, or mersagel (0·133-per-cent phenyl mercuric acetate).

Ear Canker

This is the commonest form of mange in the domestic rabbit.

Cause

Two species of mites, *Psoroptes communis* and *Chorioptes cuniculi.*

Clinical symptoms

Brownish scabs inside the ear cause repeated scratching with the hind claws.

Diagnosis

The mites can be recognized microscopically in wet films of scab after heating with 10-per-cent solution of caustic-potash solution. *Psoroptes communis* has cup-shaped suckers on segmented stalks while *Chorioptes Communis* has unstalked suckers.

Control measures

The hutches in which affected rabbits have been isolated must be thoroughly disinfected. The condition is easily cleared up by softening the scabs with a vegetable oil, removing them with forceps, and applying an efficient mange dressing such as Gamma BHC to the affected skin.

Body Mange

This is a comparatively rare type of mange in the domesticated rabbit. *Cause*

Two species of mites, *Sarcoptes cuniculi* and *Notoedres cuniculi.*
Clinical symptoms

The fur falls off from areas on the face, head, and roots of the ears, which become covered with yellowish or greyish scabs. The intense itching resulting from the infestation of the skin causes continual tearing at the affected regions with the claws.

Diagnosis

Microscopic examination of wet smears of scab preparations treated with hot 10-percent solution of caustic-potash solution serves to demonstrate the mites. Both species of mites are identified by the cup-shaped suckers on unsegmented stalks and there is no need to distinguish them, since the diseases caused by them are similar.

Control measures

The extremely contagious nature of the condition necessitates the immediate isolation of affected rabbits and the careful disposal of fatal cases by burning. The blowlamp is the best means of disinfecting contaminated hutches. Innes (1945-1946) states that Tetmosol, benzyl benzoate, and Gamma BHC will almost clear up this condition.

Internal Parasitic Conditions

Parasitic Gastritis

This condition is very rarely encountered where adequate hygienic measures are practised.

Cause

A species of roundworm, *Graphidium strigosum.*

Clinical symptoms

Loss of condition with diarrhoea, anaemia and emaciation when the infestation is heavy.

Diagnosis

A significantly high egg-count in the faeces is diagnostic in the living animal. Postmortem diagnosis is based upon an actual worn-count of the contents of the stomach, together with evidence of destruction of the mucous membrane.

Control measures

An infestation is seldom serious enough to be fatal if the feeding of contaminated greenstuffs is avoided and hutches or cages are regularly and frequently cleaned. Rabbits should not be allowed to feed on grass runs where there is any possibility of contamination with animal faeces. Phenothiazine by mouth in a dose according to body-weight is suggested for treatment of affected rabbits.

Bladder Worms

This is a parasitic condition which is relatively common in the domestic rabbit though very rarely lethal.

Cause

Cysticercus pisiformis, the larval stage of *Taenia pisiformis,* a tapeworm occurring in the dog; and also *Coenurus serialis,* the larval stage of the dog tapeworm, *Taenia serialis.*

Clinical symptoms

General unthriftiness in young rabbits results from severe infestation with *C. pisiformis.* Older rabbits may show abdominal distension due to a large number of fully-grown tapeworm cysts in the abdominal cavity.

Post-mortem examination of rabbits suffering from a severe infestation shows numerous fibrous tracks throughout the liver substance caused by the tapeworm larvae in the course of their migration into the peritoneum, where they settle down in the encysted state.

Coenurus serialis cysts occur singly in a subcutaneous position and they may reach the size of a tennis-ball. The host remains in good condition.

Diagnosis

The first condition is recognized at autopsy by the presence of numerous healed larval tracks in the liver and of a large number of the cysts in the abdominal cavity. The second is characterized by a large, rounded, soft swelling which moves relatively freely under the skin.

Control measures

The feeding of grass or other greenstuff to which dogs have had access should be avoided, as well as the contamination of foodstuff or drinking-water with dog faeces. A *Cs pisiformis* infestation is seldom sufficiently heavy to be fatal unless it follows the ingestion of grass soiled by a whole tapeworm segment excreted in a dog faex.

Coenurus serialis cysts can be removed surgically without harming the rabbit, provided they are enucleated from the subcutaneous tissue without rupture and consequent liberation of their contents.

Shackle (1944) records that the only case of residual paralysis from spinal anaesthesia he has encountered in the rabbit was in an animal with a *Coenurus* encysted in the brain.

Myxomatosis

This condition was first described in domesticated rabbits in Montevideo by Sanarelli (1893) as a highly fatal, infectious disease, which appeared to be enzootic in the indigenous wild South American rabbits, causing a comparatively low mortality in species of the genus Sylvilagus.

Later, attempts were made in Australia to reduce the wild rabbit population by calculated dissemination of the disease with generally successful results, except for the apparent development of a rabbit strain resistant to the disease. In 1953, after its introduction into France, the disease attained the scale of an epizootic throughout European wild rabbits (Or *vctolagus cuniculus),* causing almost 100-per-cent mortality and attacking domestic rabbits as well.

During October, 1953, the disease was reported among wild rabbits in Kent and Sussex. It had almost certainly come from France and it spread rapidly throughout the wild rabbit population of the Home Counties and other parts of the U.K. A detailed description of its spread is given by Thompson and Worden (1956).

The disease is readily communicable to laboratory rabbits, but it has not been conclusively transmitted to hares, mice, rats or guinea-pigs, although the natural disease has been reported as occurring in the blue (mountain) hare in Ireland.

Cause

This is a filtrable virus which has proved to be antigenically related to the virus identified as the cause of infectious fibromatosis in rabbits by Shope (1935). Transmission has been shown to occur mechanically by insect vectors, mainly mosquitoes in Australia but fleas in Britain, and also by direct and intermediate contact.

Clinical symptoms

After an incubation period varying between two to eight days, a severe blepharoconjunctivitis develops accompanied by marked thickening and sealing of the eyelids with a tenacious inspissated purulent exudate, followed by swellings on the nose and muzzle, base of ears and mucous membrane of the anal and genital openings with orchitis in the male.

According to Martin (1936), the appearance of a number of discrete, small, subcutaneous tumours, of a gelatinous and vascular nature, over the body and particularly on the ears, depends on the degree of handling of the affected rabbit and occurs in wild rabbits in only a minority of cases.

He believes that it is the small capillary lesions in the situations where a rabbit is usually seized that cause the development of skin tumours. Death, preceded by rapid emaciation and terminal pyrexia, supervenes within two to twelve days after the onset of symptoms, and in the rare instances of natural recovery permanent active immunity is established.

Diagnosis

Clinical symptoms are usually suffrcient for diagnosis. At post-mortem examination in the minority of wild rabbits which have died of the disease, but more frequently in domestic and laboratory rabbits, the characteristic small subcutaneous tumours are present around the eyes, in the peri-anal region, on the ears, and along the back.

The viscera usually appear normal although, infrequently, the spleen and lymph nodes may be visibly enlarged and there maybe oedema and petechiation of the tunica vaginalis, and the testes in the male.

The most reliable diagnostic methods are based on histopathological demonstration of the myxomatous character of the tumours in the skin or other tissues and the transmission of the disease by biological test in the laboratory rabbit.

Very suggestive features of a myxomatosis outbreak are the very high mortality and the rapidity of spread. The explosive potentialities of this infection have been well illustrated by a description of its spread in the rabbit population of Australia.

Prolonged spells of unseasonable, mild, wet weather may lead to unusually high mortality among wild rabbits owing to other conditions which may simulate myxomatosis in their high incidence and heavy fatalities.

Septicaemic pasteurellosis, acute intestinal coccidiosis, and parasitic gastritis (due to *Graphidium strigosum)* are examples, but these are easily differentiated by the appropriate bacteriological and parasitological examinations.

Treatment

Experiments on the value of vaccination and antibiotic treatment made in America, where myxomatosis has caused some losses in commercial rabbitries in isolated outbreaks, have shown that vaccines prepared from myxoma virus by treatment with heat or chemicals do not provoke immunity and that penicillin, aureomycin, chloromycetin and terramycin have no curative properties.

The disease is highly fatal, and the only practical control measure is to eliminate affected animals and to vaccinate with living Shope's fibroma virus all contact rabbits which remain without symptoms of infection throughout a period of isolation for seven days.

Prophylaxis

Myxomatosis will readily spread to laboratory rabbits unless precautions are taken to deal with it. There is little chance of spread by flying insects during the winter months but spread at this time by other means and vectors is possible and it is very likely to flare up in the spring and summer.

Control of vectors

Wherever possible, vectors should be prevented from entering rabbits' houses. This will involve proofing windows, ventilators and other openings (a wire mesh 16 to the inch-6 per cm-is suitable); the provision of air-lock doors; and the use of insecticides within the house, for example by means of aerosols. To be effective these precautions must be very thorough, and among other things call for a high standard of discipline in animal-house personnel.

Control of cases and contacts

Immediate killing of cases in the early stage of blepharoconjunctivitis, burning the carcasses, and disinfection of the cages are essential precautions. Contacts should be isolated for seven days. Wholesale destruction of contacts, even of cage mates, is not necessary.

Immunization

It has been reported that vaccination with the Boerlarge strain of the Shope (1935) fibroma virus confers a strong immunity to myxomatosis lasting for at least twelve months.

The vaccine is injected intradermally or subcutaneously: A soft fibromatous swelling appears in about five days, reaches its maximum at up to twenty-one days and then regresses, disappearing finally in one or two months.

Vaccination is ineffective in rabbits already infected with myxomatosis. Martin (1936) also found that inoculation with living Shope's fibroma virus, which produced a local tumour of short duration, gave considerable protection against myxoma virus and the Pasteur Institute. Paris, is now issuing this virus as a vaccine in France.

A vaccine is available from Burroughs Wellcome Ltd. It confers immunity for a minimum of six months on a minimum of two-thirds immunized stock. It is injected subcutaneously at a point over the shoulder. It is suggested that some system of marking rabbits which have been vaccinated should be adopted.

Listerellosis

This was first reported in epizootic form among young rabbits and occasionally guinea-pigs at the Field Laboratories, Cambridge, by Murray *et al.* (1926). Another case from the same centre was described by Paterson (1940) in a young adult pregnant female Copenhagen.

This animal did not litter down at the proper time and four days later refused food. It was destroyed when severely ill 48 hours later, the pathological features including an enlarged and friable liver studded with greyish-white pin-head foci not sharply demarcated from the normal tissue, local peritonitis, and an enlarged uterus containing dehydrated foetuses (developmental age approximately sixteen days) enveloped in a continuous circular layer of inspissated pus which could only be separated from the uterine wall with difficulty.

Profuse pure cultures of *Listerella monocytogenes* were obtained from the uterine pus, spleen and liver and from the abdominal cavity of one of the foetuses. Heart-blood cultures were sterile. Some further notes on the disease are given by Blount (1945).

Rabbit Pox

This highly contagious and fatal disease has been reported from stocks of rabbits in the Rockefeller Institute and other laboratories in the U.S.A. In these outbreaks the incubation period was five to eleven days.

The symptoms varied with the situation of the pock-like lesions but included macular and, later, papular eruption of the skin (especially on the ears) and mucous membranes (including those of the mouth and tongue), high temperature, prostration, blepharitis, enlargement and induration of lymph nodes, and orchitis. Small nodular lesions were present also in the liver, spleen and lungs.

In one extensive epidemic the mortality rate amongst young animals was 71 per cent, being highest in newly-weaned stock, while in adults it was only 14 per cent. Amongst adults, pregnant and lactating does were the most severely affected.

After recovery no carriers remained. Immunity lasted for several months and there was evidence of the transference of passive immunity from immune doe to young.

Oral Papillomatosis

Parsons and Kidd (1936) reported the existence of a virus disease causing papillomata on the mucous membranes and usually on the under-surface of the tongue. The papillomata were discrete, sessile or pedunculated, filiform or fungiform projections and were epitheliomatous in structure.

Toxoplasmosis

Toxoplasmosis has been reported.

Hydrocephalus

Hyde (1940) reported an epidemic of hydrocephalus among experimental rabbits, but was unable to ascertain the cause.

Thrombo-endocarditis, arterio-sclerosis and other diseases of the cardio-vascular system

Andrei and Ravenna (1938) reported the experimental production of thrombo-endocarditis in rabbits by the injection of various preparations of serum and milk.

They also record, however, that the condition may occur spontaneously in a small percentage of rabbits. Spontaneous sclerosis of the aorta and other large arteries is also encountered occasionally and may mislead experimentalists.

The use of the rabbit for experiments involving attempts to demonstrate the transmission of infective agents associated with cardio-vascular disease has not always yielded satisfactory results on account of these spontaneous conditions.

Sterility associated with Pasteurella septica and a Brucella bronchiseptica-like organism

Kyaw (1945) has reported the association of *Pasteurella septica* and a *Brucella bronchiseptica-like* organism with sterility of rabbits kept at the Animal Research Station, Cambridge. Breeding troubles had been encountered in these strains, affected does on postmortem examination showing enlarged uteri filled with a cheesy, sticky pus.

There had been a severe outbreak of snuffles in the stock some years previously. *Pasteurella septica* was recovered from eight out of the ten does selected from those which had bred irregularly or become sterile, and the *Bruce/la bronchiseptica-like* organism from the other two. *Pasteure/a septica* was recovered in penis swabs from four, and both organisms from the fifth, of five bucks with poor fertility records. Evidence was obtained from transmission experiments that *Pasteurella septica* could be conveyed into the uterus from the vagina by spermatozoa.

Acute Haemorrhagic Typhilitis

Innes (1945-1946) reports this as a frequent post-mortem record and suggests that *Clostridium welchii* may be responsible. Blount (1945) maintains that 90 per cent of all cases of diarrhoea in the rabbit can be attributed to non-parasitic disorders of the caecum, whereas disorders of the small intestine and liver, including all forms of coccidiosis, do not cause diarrhoea.

He says that typhilitis is most severe in half-grown rabbits, although it occurs at any age and that there is a definite seasonal incidence, with most cases occurring in the early months of the year and fewest in the summer months. There is said to be rapid loss of condition and death with a few days. On post-mortem examination haemorrhagic, anaemic and oedematous forms are described.

Appendicitis

Inflammation of the appendix is not common in the rabbit although, as Baker (1944) and other workers have noted, there appears to be a normal absorption of bacteria through the wall of this organ. Blount (1945) states that pseudo-tuberculosis is the commonest cause and he describes also a second type, to which he gives the name 'follicular appendicitis', consisting of very pate pin-head lesions lying beneath the peritoneum.

Impaction of the Colon

This condition is popularly termed 'the blows', since excessive gas formation results from the obstruction and is noted by Innes (1946) as a frequent cause of death in stock rabbits. Blount (1945) states that it occurs most commonly in half-grown rabbits during the summer months of the year, and his own cases have been associated with the feeding of clover.

The post-mortem findings recorded by him include hardened dry and firm caecal contents, with faecal matter adherent to the lining membrane in severe cases, differential states of activity of parts of the intestines, excess of gas in the stomach in some cases, and congestion of the lungs attributed to mechanical pressure.

There would appear to be analogies between this condition and certain disorders of cattle associated with clover feeding.

Subcutaneous Abscesses

Blount (1945) considers that where large numbers of rabbits are kept about 2 per cent may be expected to die, or to have to be killed, as a result of local injuries with consequent abscess-formation.

These injuries he attributes mainly to fighting, and in fatal cases he records that the liver is often degenerate, brown, and with feathery markings.

He advises early separation of the sexes, separation of bucks which have been in service from unmated bucks, the maintenance of small units, and the avoidance of wire mesh in breeding compartments.

Torticollis

Innes (1946) records this as a frequent post-mortem record in stock rabbits. Blount (1945) provides an illustration of a case but does not appear to regard it as a common condition.

Spontaneous Paralysis of the Hind Quarters

Jabotinsky (1936) recorded the histological findings in the tissues of three cases which had shown spontaneous paralysis of the hind quarters and for which no satisfactory explanation had been offered.

Details were given of the multiple necrotic foci scattered throughout the grey and white matter of the lumbar section of the spinal cord. Similar changes were seen in the nerve ganglia. No conclusions were reached concerning aetiology.

Injuries

Injuries such as a broken back and broken leg are not uncommon among laboratory rabbits and may escape accurate diagnosis. A careful examination for such injuries should always be made.

Blount (1945) reports the occurrence of broken back in does and attributes it to a sudden leaping up from fright while the animal is sitting partly in and partly out of the sleeping compartment, with consequent rupture of the lumbar region.

INDEX

D

R

S

T

U

V

W

X

Z